DRYWALL CONTRACTING

by

JAMES T. FRANE

CRAFTSMAN BOOK COMPANY
6058 Corte del Cedro, Carlsbad, CA 92009

To Tré, Sara and Bryan for their encouragement and help.

Library of Congress Cataloging-in-Publication Data

Frane, James T.
Drywall contracting.

Includes index.
1. Dry wall--Contracts and specifications. I. Title.
TH2239.F73 1987 690'.12 87-24303
ISBN 0-934041-26-1

Photos by Tré Frane
Illustrations by the author
Graphic design by Marimi De La Fuente
Second printing 1990

CONTENTS

Chapter 1

Chapter 2

Chapter 3

Chapter 4

Chapter 5

Chapter 6

Chapter 7

Chapter 8

Chapter 9

Chapter 1

DRYWALL APPLICATIONS

Gypsum has been used in construction for thousands of years because it's cheap, durable and easy to work with. In the 1940's and 1950's, gypsum drywall replaced plaster as the material of choice for interior wall covering. Before inexpensive, high quality drywall was available, interior walls in most homes were lathed and plastered.

Plaster is still an excellent wall cover. But application takes considerably more labor, the work must be done by specialists, it's messy, and it leaves the building saturated with moisture that has to evaporate before painting and finishing can begin. Gypsum drywall can be hung by carpenters, it costs less, leaves less of a mess, and requires far less moisture for application. That's how gypsum wallboard got its name, *drywall*. The wall is covered with a dry material that can be finished shortly after application is complete.

You may know gypsum drywall by other names. For example, some builders call it *sheetrock* or *plasterboard* or *gypsum wallboard* or simply *gypboard*. All these names refer to drywall.

To simplify matters, we'll call the product *drywall* throughout this book.

If you make your living with drywall, there are some basic principles you should understand. That's the subject of this first chapter. I'll describe how drywall panels are made. Then I'll suggest some advantages you may not have considered in using drywall. We'll discuss common panel sizes and types. Finally, we'll take a brief look at the various panel edge treatments.

From Mineral Deposit to Drywall

Gypsum is found all over the world in natural deposits that can be mined. In its purest form, gypsum is white. But in the earth it can be different shades of gray, brown or even pink, depending on the impurities present. These impurities include clay and iron oxides.

Here's how the raw mineral is converted into those familiar white panels. Once the mineral is removed from the earth, it's crushed into small particles. These particles are then heated to about 350 degrees Fahrenheit to drive off nearly all the moisture. This turns the gypsum into a dry powder, which is then mixed with additives, aggregates and fibers to add strength and moisture resistance.

Next, water is added to this mixture so it can be molded and shaped into the form needed. For drywall, the moist gypsum is formed between two layers of paper or other covering. As the gypsum mixture dries, crystals form and interlock, gradually turning the board into a rock-hard mass. Before it's hardened completely, the board is cut to length and passed through dryers that remove any free moisture. This is important. Drywall swells when it gets wet and shrinks as it dries. Too much shrinking and swelling will spoil any drywall job.

ADVANTAGES OF DRYWALL CONSTRUCTION

Drywall is cheap compared to other wall materials. But there are other advantages: fire protection, noise insulation and ease of installation.

Superior fire protection: When drywall is exposed to high temperatures, as in a fire, here's what happens. The outer layer of gypsum that's exposed to heat releases water in the form of steam. This has a cooling effect and limits the temperature rise in the drywall. This process is known as *calcining*. When calcining takes place, the dehydrated gypsum turns back into a powder that's a good insulator. Heat at the surface is kept away from gypsum below the surface.

Noise protection: The relatively high density of gypsum provides better sound damping than lighter wall materials such as plywood or hardboard. The more dense a material is, the greater its sound-absorbing capacity.

Ease of installation: Drywall panels are easy to install. After World War II, some building materials weren't available and there was a shortage of skilled tradesmen. Builders were eager to try materials that were available and could be installed quickly. Drywall became the most popular wall covering material in nearly every community in the U.S.

PANEL SIZES

Drywall panels are normally 4 feet wide and 8, 10, 12, 14, or 16 feet long. The first drywall panels were sold as small, 3/8-inch thick sheets to be used as a lath backing for plaster in remodeling work. Now there are six standard thicknesses available: 1/4-inch, 5/16-inch, 3/8-inch, 1/2-inch, 5/8-inch and 1-inch.

1/4-inch: Use this thickness to cover existing ceilings and walls where you don't want to remove the original paneling or plaster. You can also use it in multi-layer installations to increase the noise insulation in a wall or ceiling.

5/16-inch: This thickness is relatively lightweight. But you can use it for single-ply installations if the framing members are spaced correctly. This thickness is used in mobile homes to keep weight to a minimum.

3/8-inch: Like the 1/4-inch thick panels, you can use 3/8-inch panels to cover existing wall and ceiling surfaces. You can also use them in multi-ply systems.

1/2-inch: Use this thickness for single-ply wall and ceiling construction in stationary buildings. You can also use it in multi-ply installations to increase the fire resistance rating or improve sound control.

5/8-inch: This thickness provides an increased fire code rating and greater sound insulation. It's more rigid than 1/2-inch drywall so it offers more resistance to sag. Use 5/8-inch thickness for single-ply construction where there are large, uninterrupted wall or ceiling areas. The 5/8-inch thick panels can be used on longer spans between joists and between studs.

1-inch: These panels, known as *coreboard*, come two ways: as a single 1-inch thick panel or as two 1/2-inch thick panels that have been laminated together. Use 1-inch thick panels in solid drywall partitions or in installations that require a high fire code rating.

PANEL TYPES

Most interior rooms in a house or office building will take standard drywall panels. But some structures have special requirements, such as increased fire resistance, water resistance, exterior applications, backing board, coreboard, formboard, lath, blocking, radiant heating, decorated and veneer-base panels. Let's look at these special panels.

Standard Drywall Panels

Standard drywall panels have a paper covering on each side and on the long edges. The backs of the panels are surfaced with gray liner paper. The facing is a manila paper which extends over the long edges. The manila facing paper is also light gray. It's smooth and will take a wide variety of finishes.

The long edges are usually tapered, as shown in Figure 1-1. This is a shallow taper about 2½ inches wide. It reduces the edge thickness by only about 1/16 inch. The taper makes it easier to tape and fill the joint between two pieces of drywall. When filled, taped and sanded, the surface at the joint should be as smooth and even as the board itself.

Standard drywall is available in thicknesses of 1/4, 5/16, 3/8, 1/2 and 5/8 inch and is 4 feet wide. Most dealers stock the 8-foot length. But lengths up to 16 feet are available in some thicknesses.

The length of panel you use will depend on the application and panel thickness required. Longer sheets are heavier and harder to handle. They're also more likely to break during handling. You have to be careful with them. But longer sheets can cut down on the number of joints you have to finish. Weigh the advantages of longer sheets with fewer joints against the inconveniences that come with longer panels.

Use standard drywall where there are no special requirements. You can use it in most rooms in a house or office building. Where standard drywall isn't good enough, select the special panel you need. The next few paragraphs describe special applications.

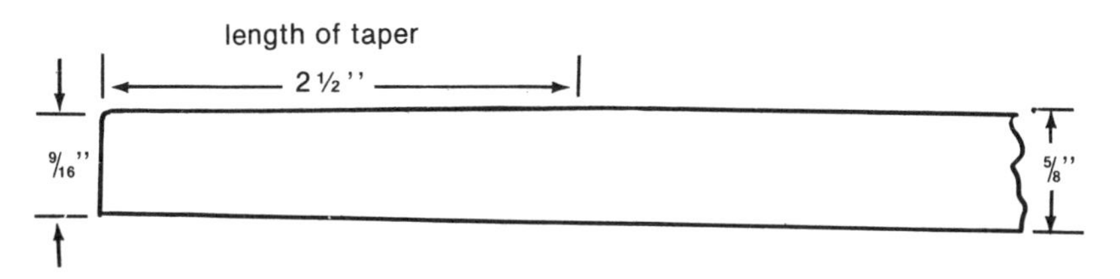

Cross section of tapered edge
Figure 1-1

Fire-Resistant Panels

Fire-resistant drywall panels are usually marked *Type X*. Type X panels look like standard drywall panels except that they're stamped "Fire-Resistant" or "Type X" on the back of the panel.

Drywall already has its own built-in fire protection. In Type X panels, the fire resistance is increased by adding glass fibers to the core. The panels come in both 1/2- and 5/8-inch thicknesses. The 1/2-inch thickness has a fire rating of 45 minutes. The 5/8-inch thickness has a 60-minute fire rating.

Use Type X panels wherever you need high fire resistance. The extra fire protection they offer may let you use thinner drywall. For example, one thickness of Type X may give the same fire rating as multiple layers of standard drywall. To get maximum fire resistance from Type X panels, you have to follow special joint finishing procedures. Otherwise, fire and heat would go around or between the panels. I'll describe the procedure in detail later.

Fire-resistant drywall panels are available in 4-foot widths and in lengths from 8 to 12 feet.

Foil-Back Panels

Foil-back drywall is also known as *insulating drywall.* Foil-back drywall is made by laminating aluminum foil to the back surface of drywall panels. The foil creates a waterproof membrane, creating an effective vapor barrier. The reflective surface of the foil adds to the insulating value of the drywall. Foil backing is available on many types of drywall panels for different applications.

You can use foil-back panels with steel or wood framing or furring. But don't use them as a base for tile or other highly moisture-resistant coverings in hot, humid climates. If you do, the gypsum core will absorb and trap some moisture. Trapped moisture will cause the core to deteriorate, weakening the drywall. Eventually, it will begin to warp, sag or crumble.

Water-Resistant Panels

Water-resistant drywall panels are also known as *WR panels* or *greenboard.* Use these panels in areas that are exposed to high moisture. Kitchens and baths are good candidates for water-resistant panels. The facing paper on WR panels is light green. The backing paper is gray or brown.

On WR panels, both the paper covering and the core are water-resistant. The paper is multilayered and is treated with chemicals that keep moisture out. Asphalt compounds are added to the gypsum core so it won't absorb any moisture, even when there's a tear in the paper.

Water-resistant panels are also available with a high fire resistance rating. Use these panels in situations where fire separation is required but where the panels may be exposed to high moisture during normal use or during construction.

Apply WR panels directly to the studs without a vapor barrier behind them. Be absolutely sure you install the panels with water-resistant joint compound and finish them according to the manufacturer's recommendations.

WR panels come 4 feet wide and 8 to 12 feet long. They're available in 1/2- and 5/8-inch thicknesses. Fire-resistant WR panels are 5/8 inch thick.

Exterior Ceiling Panels

Use exterior drywall ceiling panels on horizontal overhead surfaces, such as soffits, canopies and carport ceilings. These panels are weather-resistant as long as they're not directly exposed to the weather. A water-resistant gypsum core is wrapped with water-repellent beige facing paper suitable for decorating. Finish the joints with battens or joint tape and water-resistant joint compound.

Exterior ceiling panels come 4 feet wide and 8 to 12 feet long. The panels are available in 1/2- and 5/8-inch thicknesses. The 5/8-inch panels are fire-code rated.

Exterior Sheathing

Exterior drywall sheathing panels are designed for indirect exposure to the weather and installation on vertical surfaces. Their water-resistant core is covered by tightly bonded, water-repellent paper. Using these panels eliminates the need for sheathing paper. They also add to the fire resistance of the structure. Apply the sheathing panels directly to the framing members of a building.

Exterior drywall sheathing must be covered with an exterior finish. Aluminum, stucco, masonry and shingles are all suitable finishes. Be sure to attach the finish to the framing members, not just the sheathing.

Exterior sheathing panels come in 5/8- and 1/2-inch thicknesses. The 5/8-inch thickness is fire-code rated. The 1/2-inch panels are available in 2-foot widths that have V-shaped, tongue-and-groove edges, as shown in Figure 1-2. Apply these panels *horizontally* on *vertical* surfaces. The V-grooves aid in the shedding of water.

Both the 1/2- and 5/8-inch thicknesses come 4 feet wide and from 8 to 12 feet long. The 4-foot wide panels normally have square edges. You can install these panels with the long edges vertical or horizontal.

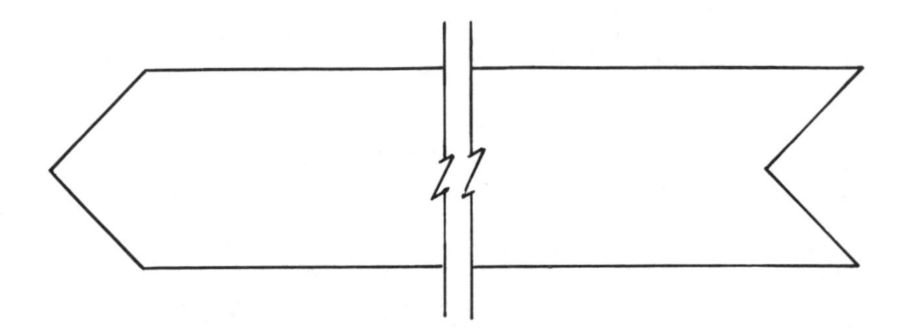

V-shaped tongue and groove edges
Figure 1-2

Backing Board

Gypsum backing board is designed to serve as the first layer in a multilayer gypsum wall or ceiling. Both surfaces of backing board are covered with liner paper and are not suitable for a decorative finish.

Backing board is also available with a foil backing. And there's also a moisture-resistant backing board that has a vinyl covering. This is intended for high-moisture areas, such as showers, where tile is applied over the backing surface. You must seal the cut edges of backing board to keep it water resistant.

Backing board is 4 feet wide and 8 to 12 feet long. It's available in 1/2- and 5/8-inch thicknesses. The 5/8-inch thick panels are fire-code rated.

Gypsum Coreboard

Gypsum coreboard is 1 inch thick and is designed for solid gypsum partitions without framing members. It consists of two factory-laminated layers of 1/2-inch gypsum backing board. Coreboard is manufactured with V-shaped, tongue-and-groove longitudinal edges for use in solid partitions. See Figure 1-2.

If you're putting up partitions with spaced core sections, order coreboard with square edges. Square-edge coreboard comes scored at 6- or 8-inch intervals so it breaks easily into sections.

Gypsum Formboard

Gypsum formboard panels are used as the form for poured gypsum concrete roof decks. Leave the formboards in place after the gypsum concrete sets up. The formboard serves as the ceiling surface in the finished installation. Gypsum formboards are faced either with manila paper suitable for decorating or with vinyl that needs no other finishing.

Fasten the formboards to the beams or joists and then pour gypsum concrete into the form. The combination of formboards and gypsum concrete creates a noncombustible roof. The formboards aren't considered structural members for purposes of calculating the load-carrying capacity of the completed roof. Formboards are 1 inch thick, 8, 10 or 12 feet long, and come in widths of 24, 32 and 48 inches.

Gypsum Planks

Gypsum planks have steel reinforcement for use as structural roofing members. A galvanized wire mat is cast into the planks, and the planks have galvanized steel tongue and groove edging. They're not water-resistant, and so should be covered with roofing as soon as possible after installation. Gypsum planks are 2 inches thick, 15 inches wide and 10 feet long.

Gypsum Lath

Gypsum lath can be used as the base for plaster on walls and ceilings. It consists of a gypsum core covered with rough paper that creates a good bond with the plaster. The top layers of paper absorb moisture quickly and evenly from the plaster, so that it will set up properly. The paper near the gypsum core is moisture resistant to protect the core. Installation is quick, and it provides a smooth, continuous surface for the plaster.

Lath is available with a solid face or a perforated face. The perforated face has 3/4-inch holes drilled through the board at 4-inch intervals. When you use perforated lath, the plaster is forced through the holes. This increases the bond between the plaster and the board. It also increases the fire resistance.

Gypsum lath is available 3/8-inch and 1/2-inch thick, 16 and 24 inches wide, and in lengths of 4 feet and 8 feet.

Partition Blocks

Solid gypsum partitions use partition blocks rather than studs for framing members. Partition blocks come in 2- to 6-inch thicknesses. The precast blocks are 12 inches high and 30 inches long. These blocks are made from a mixture of gypsum and fibers. Fibers are added to increase the block strength. Laminate standard gypsum wallboard to both sides of the partition blocks to form the finished wall or partition.

Radiant Heating Panels

Gypsum radiant heating panels have electric resistance heating elements embedded in the core of each panel. You can install these panels in suspended ceilings or use nails or screws to fasten them directly to joists. When using radiant heating panels in walls, fasten the panels directly to the studs.

The panels are 5/8-inch thick and have long electrical leads that connect to a circuit by means of a junction box. The panels come 4 feet wide and are 4, 8 or 12 feet long.

Decorated Drywall

Decorated drywall is standard drywall covered with decorative paper or vinyl. It's available in a variety of colors, patterns and finishes. You can install it with color-matched nails or adhesive. It requires no further finishing after installation.

The panel edges are either tapered or square and are butted together. It's harder to make good joints between these panels. That's why you won't use decorated panels for ceiling applications. Sometimes special moldings are used to conceal the joints.

Matching or contrasting moldings are available for decorated drywall panels. Use these moldings to cover and protect corners and panel joints. Moldings help protect the panels from damage.

Veneer-Base Drywall

Use gypsum veneer-base drywall when you want to apply a veneer plaster coating to the surface. Veneer plaster is suitable for all interior applications.

Veneer finishes are durable and wear-resistant. Use them in the heavy-traffic areas of a building. You can apply the veneer coating in one or two layers. You'll apply it 1/16 to 3/32 inch thick, directly to the drywall. The finish can be either smooth or textured.

Gypsum veneer-base drywall is covered with multiple layers of paper that provide strong adhesion to the veneer plaster. The outer layers of paper absorb the moisture from the plaster, ensuring a strong bond. The inner layers of paper are moisture-resistant to keep the gypsum core dry.

Veneer-base drywall comes in 4-foot widths with square-cut edges. Standard veneer-base drywall is 3/8-inch, 1/2-inch or 5/8-inch thick. Type X (fire-code rated) is only available in widths of 1/2 inch and 5/8 inch. Both types come in 8, 10 and 12-foot lengths. Install it with nails, screws, or adhesive.

PANEL EDGES

So far we've concentrated on panel core compositions and surface coverings. But that isn't the whole story. Now let's take a look at the panel edges.

Unfinished panel edges need special treatment. Exposed, unfinished panel edges should have trim or corner reinforcement installed. Butted panel edges may require expansion joints to prevent buckling in warm and humid weather. Let's look at each of these panel edge treatments.

Trim

Exposed, unfinished panel edges are common at window and door openings. The two most frequently used panel edge trims are metal trim and vinyl trim.

Metal trim— This comes in U and L shapes. Apply it to exposed edges of drywall to protect it and give a finished appearance. Fasten the trim to the drywall by nailing through the flange of the trim. Flanges are available in solid metal or in mesh. Cover the flanges with joint compound or veneer plaster.

Vinyl trim— Apply vinyl trim to drywall edges and intersections. You can use it as a decorative finish for exposed edges. Or you can use it as a flexible seal in place of caulk. It can also provide stress relief at the edges of panels.

Vinyl trim comes in a variety of colors and can be painted. There's even vinyl trim that looks like wood. It's more economical than clear grade wood. And it's more resistant to damage and wear.

Corner Reinforcement

Corner reinforcement, also known as *corner bead*, protects external corners from damage. Apply it to external corners that are subject to wear.

Corner bead is made of metal, so it's stronger than the edges of the drywall. You can buy it in solid galvanized steel or wire-mesh galvanized steel. It's V-shaped and has an angle of slightly less than 90 degrees. A raised section, or bead, at the bend of the "V" provides protection against impact damage. The bead also serves as a screed or guide for the drywall knife, aiding in the application of joint compound. Each half of the "V," called a *flange*, is from 1 to 1¼ inches wide, depending on the particular corner bead.

You can nail or staple the corner bead into place. Or you can use a special clinching tool. Use wire mesh corner bead with veneer plaster. Use the solid steel corner bead with all types of drywall construction.

Expansion Joints

Where there are large expanses of ceiling or wall, use expansion joints to relieve the stress caused by expansion and contraction. On ceilings, install expansion joints from wall to wall. On walls, install them from door header to ceiling or from floor to ceiling.

Most expansion joints are made of formed zinc. They fit between the drywall panels. The center section of the expansion joint is V-shaped. The "V" is flanked by two flanges. Cover the flanges with joint compound or veneer plaster.

For radiant heating systems, there's an expansion joint with flanges that fit behind the panels. When you use expansion joints in systems that have special sound control requirements, you'll have to install seals behind the joints.

That's about all you need to know about drywall materials. There's no need for you to memorize the information in this chapter. Just remember where you read it so you can look up what you need to know when a question comes up on some job. In the next chapter we'll take a close look at some of the specialized tools drywall hangers use to speed and simplify the work they do.

Chapter 2

TOOLS OF THE TRADE

There are three categories of tools you'll be using in drywall work: measuring tools, installation tools, and specialized tools. In this chapter, I'll cover everything you need to know about drywall tools and describe how to use them to best advantage.

MEASURING TOOLS

Accurate measurements are essential for proper fit when installing drywall panels. The measuring tools you'll need include: a tape measure, level protractor, dividers, level, square, plumb bob and chalk box.

Tape Measures

The handiest tape measure for all-around use is the spring-loaded retractable tape. It's available with either black markings on white tape or black markings on yellow. Black on yellow is usually easier on the eyes, but you won't be staring at a tape measure all day long. So choose the color combination that you find easiest to work with.

Tape measures come in different lengths and widths. A 12-foot long tape is adequate for measuring drywall — but it may not be the best choice for measuring a room. A 20- or 25-foot long tape works better. Tape widths vary from 1/4 inch to 1 inch, with most of the longer tapes being at least 3/4 inch wide. I recommend a wide tape. The narrower the tape, the more flexible it is. You need a stiff tape for drywall work. The stiffer a tape measure is, the farther you can extend it without it sagging or folding. A stiff tape lets you measure well beyond your actual reach. It also lets you take measurements by yourself where two people would normally be required.

An alternative to the retractable tape measure is the folding rule. This can be handy, especially for long reaches. The tape measure and folding rule are both shown in Figure 2-1.

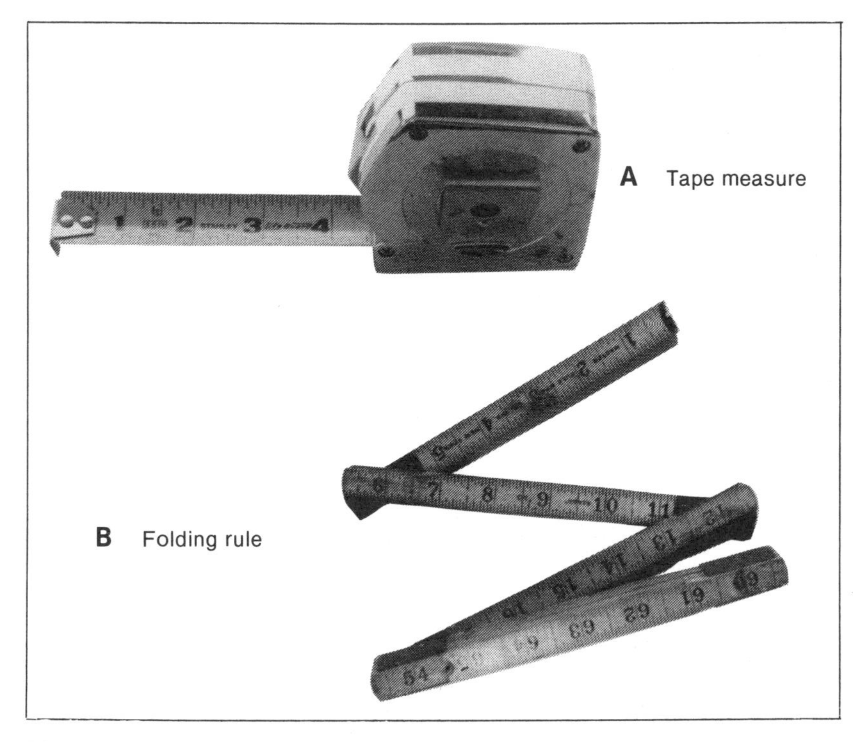

Measuring tools
Figure 2-1

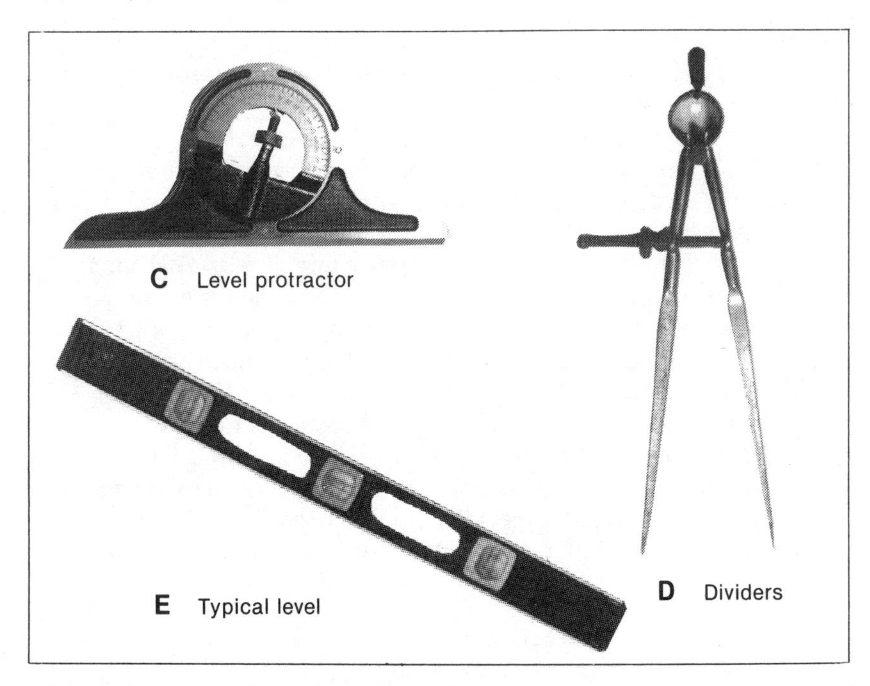

Measuring tools (continued)
Figure 2-1

Level Protractors

There will be times when you'll have to measure drywall to fit a surface that isn't at right angles (90 degrees) to the edge of the drywall. To cut the drywall to fit the space, first measure the angle of the space. There are several ways to do this.

Let's look at an example. Assume the ceiling is angled, rather than horizontal. From the highest point on the ceiling-to-wall intersection, measure down 4 feet. Using a level, draw a horizontal line from this point. Measure 8 feet horizontally from the point along this level line. Mark a new point at this location. Measure straight up from this new point, using a plumb bob or level to make sure you're going straight up.

This vertical distance will be less than the 4 feet you measured up from the first point because the ceiling is sloping down toward the horizontal line. Let's say this last vertical measurement is 3 feet. This tells you that over an 8-foot horizontal distance, the ceiling slopes 1 foot (4 feet minus 3 feet).

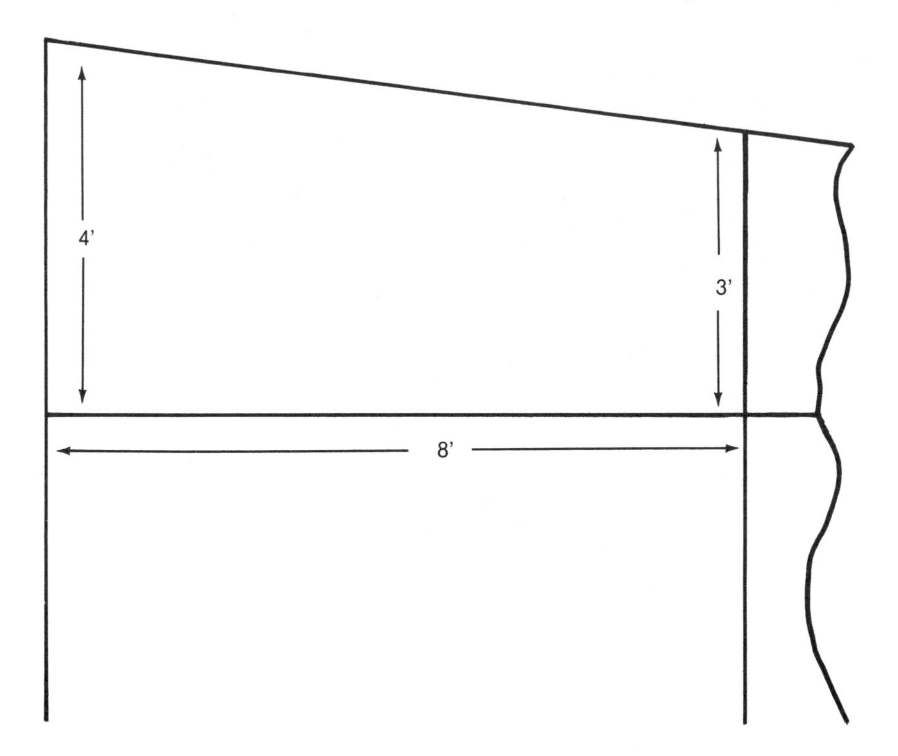

Angle cut on wallboard panel to fit sloped ceiling
Figure 2-2

On the drywall panel you're installing in this location, mark a point 3 feet up from the bottom of the panel along the end that will go where you measured 3 feet up to the ceiling. Snap a chalk line between this point and the upper corner on the other end of the panel. This will be your cut line. Once you've made the cut, the panel will butt tightly against the ceiling at the points you measure. It'll look like the panel in Figure 2-2.

This process is just the same as if there were no ceiling and you could place the drywall against the wall and scribe a line that's the same angle as the top of the wall. The only difference is that you're doing it by transferring a measurement of the angle to the drywall, instead of scribing the line in place. These instructions assume that you're installing the panels horizontally. If you're installing the panels vertically, just measure a 4-foot horizontal distance instead of an 8-foot horizontal distance. The other steps are the same.

There's another method for measuring and marking the angle to be cut. This method uses a level protractor. A protractor is a metal or plastic half-circle marked off in degrees. It's shown in section C of Figure 2-1. Here's how to use it. Place the straight section of the protractor on one edge of a panel. Make sure you place the center of the straight edge of the protractor where two panel edges come together.

The level protractor has a bubble level bisecting the circular section. This circular section can be rotated. Now rotate the circular section until the bubble in the level is centered. You can then read the angle of the edge or surface from the protractor.

Let's say the angle is 10 degrees. This means that the top edge of the panel, after it has been cut, will have to be 10 degrees from horizontal in order to fit properly against the ceiling. You really don't care what the actual angle measurement is, though. You'll be transferring the measurement using the level protractor, which you have positioned to the correct angle. No matter what the angle is, the level protractor is now locked in the proper position.

Your next step is to transfer this 10-degree angle to the drywall. Your level protractor is only a few inches long, so it won't work to scribe the entire cut line onto the drywall panel surface. It will work, however, to mark a section of that line. A level protractor has a slot in which you can fit the straight (ruler) part of a combination square. This will give you a longer reference edge for transferring your angular measurement to the drywall panel.

Lay the drywall panel flat and place the lever protractor in the corner where the cut will start. (You could do it with the panel standing on edge, but this is easier.) Position the level protractor so that the level portion of the tool is parallel to the drywall edge to be cut. The base of the level protractor will then be at the proper angle for the cut. Scribe a line on the panel surface along the base of the protractor. This line represents the angle to be cut. If this line won't intersect the corner of the drywall panel, mark a line parallel to it, using a long straightedge or a chalkline, that does intersect the corner at which the cut is to start.

Dividers

Dividers consist of two thin metal rods with points on one end and hinged together at the other to form an "A" shape.

See Figure 2-1, section D. You can move the pointed ends of the divider closer together or farther apart to match a specific dimension. Set the pointed ends at the correct dimension and transfer that distance to any other location. This simplifies repetitive measurements.

Some dividers have a screw-thread adjuster to move the points together and apart. The screw-thread adjuster allows for more precise adjustment. It also helps hold the adjustment as the dividers are moved from one place to another.

Levels

A level can also be considered a measuring device. True, it doesn't necessarily measure length (although some levels have a length scale), but it does determine when an edge or surface is level or plumb. (Plumb means straight up and down.) Use a level to check the framing before you begin installing any drywall. For drywall panels to fit properly, the framing must be level and plumb. A level is also handy when marking cutouts on drywall panels.

A level is a straight metal, plastic or wood bar, with glass or plastic tubes containing fluid. There's a small bubble in the fluid. There's usually at least one horizontal tube and one vertical tube, and there may be a 45-degree angle tube. This allows you to make a variety of measurements or to check for plumb and level. Section E in Figure 2-1 shows a typical level.

Some levels are small and light enough to mount on a string. Use a string level to adjust the height of two points that are far apart. There are also levels available up to at least 8 feet in length. For most drywall work, a 2-foot long level is enough.

If you want more than one size of level, you might select an 8 or 10-inch level for tight places and a 4-foot long level for larger areas. The 4-foot level has the advantage of being a good straightedge for drawing cutting lines and for use as a cutting guide.

There's another level that's handy for checking the level between two distant points: a length of flexible clear plastic tubing. Fill the tubing with water and hold the ends of the tubing at the two points to be checked. Keep the ends of the tubing above the water level so water doesn't run out. Stretch the tubing between the two points. It doesn't matter if the tubing is too long.

The height of the water in the tube will be exactly the same at both ends. This type of level is more accurate than mounting a level on a string stretched between two points. Even though a string level doesn't weigh much, it can still cause the string to sag a little. In fact, if you stretch a string over a great distance, the string will sag no matter how tightly you stretch it.

Squares

Use a square to mark right angles precisely. It's usually marked off in inches along one or both legs so it can be used as a ruler. Use a square to mark and check the right angles at corners, edges and surfaces. Squares come in many sizes and in fixed and adjustable types. Figure 2-3 shows three common types of squares.

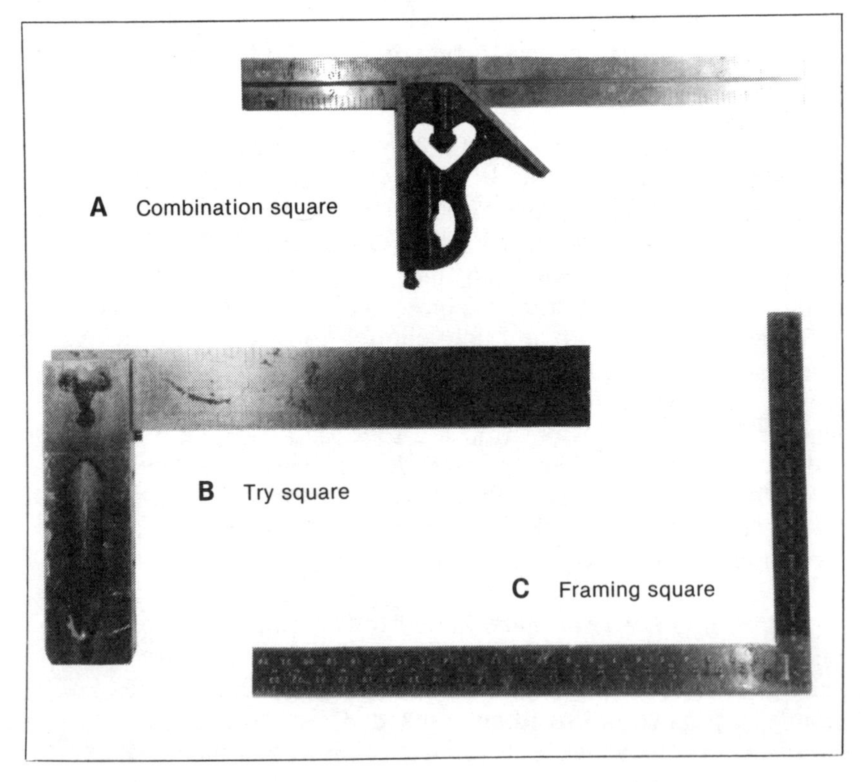

Squares
Figure 2-3

Combination square— The adjustable or combination square is most versatile because it can also be used as a depth gauge. Combination squares have a bubble level and a 45-degree straight section as part of the handle. The handle is adjustable.
If you file a notch at the center of the end of the longer leg, you can hold a pencil point there while sliding the square along an edge, marking a cutting line parallel to that edge.

Try square— Because the try square has fixed legs, it's slightly more accurate than the combination square. This degree of accuracy is more important in cabinetmaking than in drywall work.

Framing square— A framing square is larger than the combination square or the try square. Like the try square, the framing square also has fixed legs. The larger size allows you to work accurately on larger surfaces.

Plumb Bobs

Use a plumb bob to determine whether a surface or an edge is vertical, or *plumb*. Or you can use it to locate a point that is directly under another point. The bob is cone-shaped with a sharp point, as shown in Figure 2-4. It's suspended by a string attached to the blunt end of the bob.
A plumb bob used in drywall work should weigh at least 4 ounces so it stretches the string tight enough when allowed to hang free. Gravity will hold the string perfectly vertical, as long as the bob is hanging free and isn't moving.
In an emergency, you can tie a heavy nut or other object onto a string and use it as a plumb bob. The disadvantage of using a nut is that there's no sharp point to locate the precise spot under the string. But if the end spot isn't important, the nut will do the job just as well.

Chalk Boxes

Use a chalk box and chalk line to mark a straight line on a large surface. It's easier than using a straightedge, unless you're marking very short lines. You can use a chalk box on almost any surface you have to mark in drywall work.

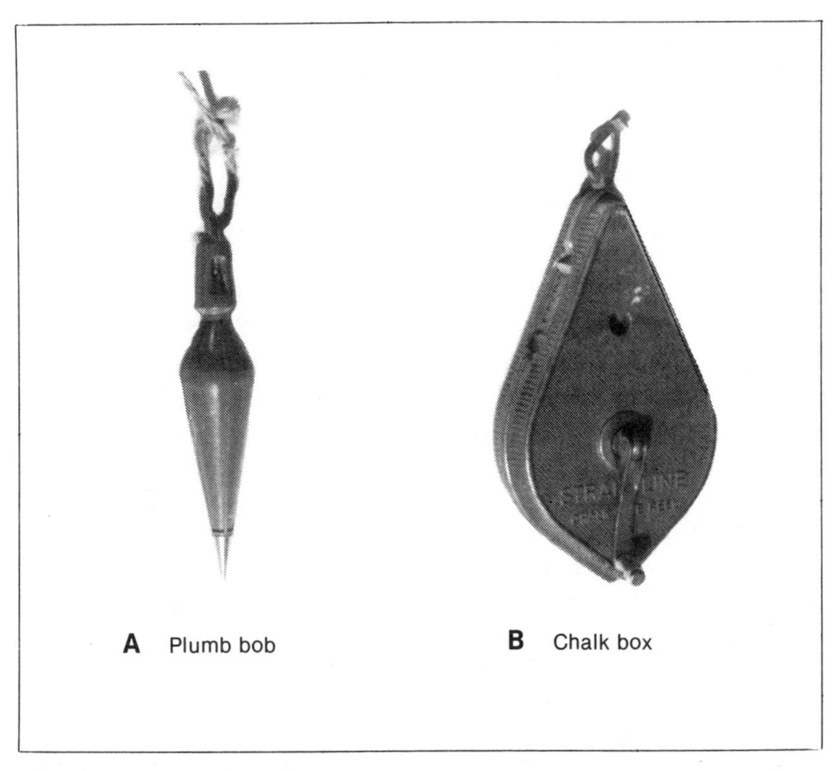

Plumb bob and chalk box
Figure 2-4

A chalk box contains powdered chalk and a string wound on a reel. The reel has a crank for rewinding. See Figure 2-4. The box is made of metal or plastic and the string is drawn out of it from one end. The other end is pointed so it can be used as a plumb bob.

The string has a hook on the end so you can attach it to the edge of a board or other surface. As you pull the chalk box, the string unwinds from the reel. The string is coated with powdered chalk. Here's how to mark a good line. First, pull the string out the required distance. Hold the chalk-box end of the string against the surface to be marked. Stretch the string taut. Then grasp the string about half way between the box and the hook. Pull the string away from the surface and let it snap back. The chalk-covered string, snapping against the surface, leaves a chalk mark running straight from the chalk box to the hook at the other end of the string.

There are a variety of colors of powdered chalk available. Blue stands out well against almost any surface. Red is available, but I don't recommend it — it's hard to cover up later. The chalk box is one of the handiest and most accurate methods of marking a straight line I know of.

INSTALLATION TOOLS

Installation tools are the tools you'll use to cut the drywall panels and fasten them in place. There are four important types of installation tools: nail drivers, screwdrivers, knives and saws.

Nail Drivers

The most common tools used for nailing drywall are the claw hammer and the drywall hammer. Both tools work equally well. Which one you choose is a matter of personal preference. Some people don't like to see the hatchet end of a drywall hammer coming toward them on every backstroke. You may also want to consider using a magnetized hammer.

Consider the weight, balance and "feel" of a hammer before making your selection. You'll be using your hammer many hours a day. A hammer that's the right weight, is properly balanced, and fits your hand well reduces strain and fatigue. A hammer that isn't well balanced requires more force both on the backstroke and on the driving stroke.

Claw hammer— A claw hammer has a round, flat face for driving nails. On the other end of the head, there's a split claw for pulling nails. See Figure 2-5.

The face of the hammer head can be either smooth or cross-hatched. A smooth face is more likely to slip off the head of the nail when you strike it. A cross-hatched face grips the head of the nail better.

The claw is either straight or curved. The curved claw provides greater leverage when pulling nails. The straight claw is more versatile for prying or wedging because you need less of an angle to fit the end of the claw between objects.

The handle may be steel, wood or fiberglass. Steel handles have rubber grips to reduce the shock to your hand. Hammers with wood handles usually cost less. The hammer head is more

A Claw hammer

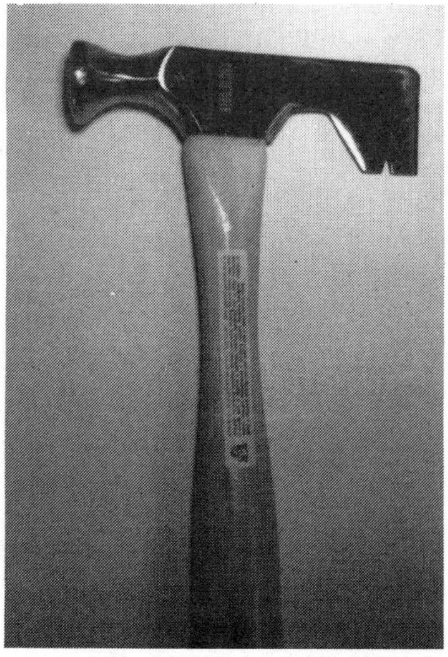

B Drywall hammer

Nail drivers
Figure 2-5

likely to loosen on a wood handle. Also, wood handles are more likely to break after extended use. But wood absorbs shock well. And many people prefer the feel and grip of wood in their hand.

Drywall hammer— The drywall hammer looks like a hatchet. See Figure 2-5). It has a round, slightly convex face. When driving a nail, the convex face makes a dimple in the surface of the drywall and drives the nail head slightly deeper into the drywall. You later fill the dimple with joint compound, concealing the nail heads. The other end of the hammer head tapers to a hatchet blade that you can use for wedging and prying. The blade isn't sharp and isn't designed for cutting.

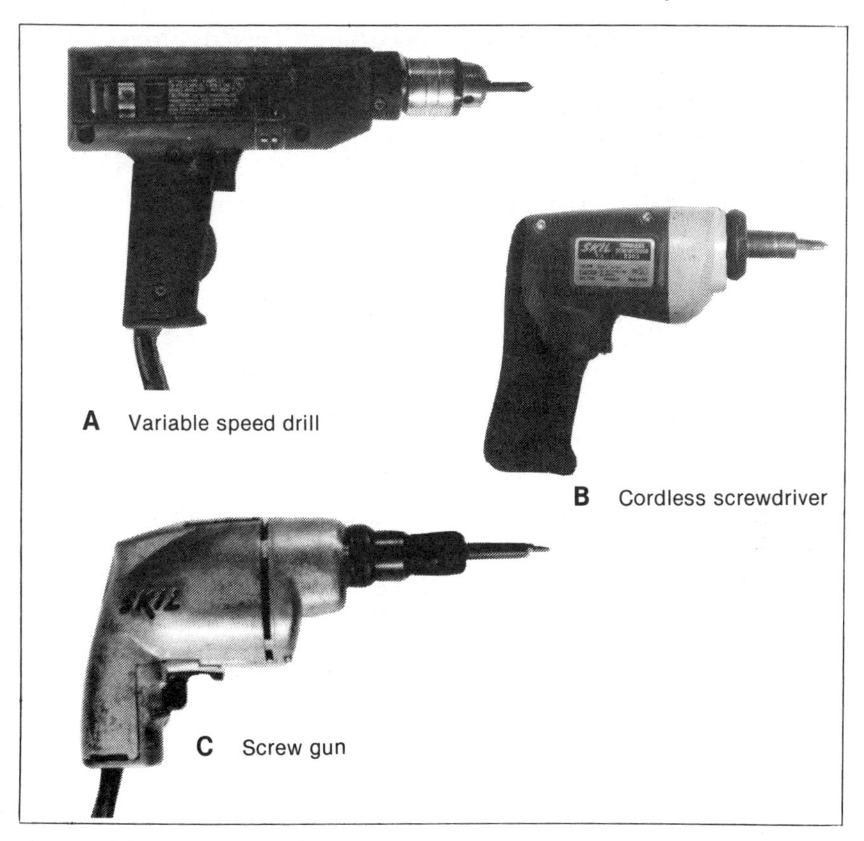

Screwdrivers
Figure 2-6

Magnetized hammer— A magnetized hammer holds the nail for you, leaving your other hand free. Driving the nail exactly where you want it takes a little practice. The magnetic portion of the hammer head will eventually lose its magnetism, but it's replaceable.

Screwdrivers

You could use a manual screwdriver for installing drywall, but it's slow and takes too much effort to be practical. Power screwdrivers are the only way to go. There are three common types: the variable-speed electric drill, the electric screwdriver and the electric screw gun. Figure 2-6 shows all three.

Variable-speed electric drill— This drill uses a screwdriver bit in the chuck.

Electric screwdriver— Most electric screwdrivers turn at a constant 300-rpm speed. Some have a 600-rpm speed. Figure 2-6 shows a cordless version.

Electric screw gun— The electric screw gun has a clutch that causes the screwdriver bit to turn when pressure is applied. It also has a depth adjustment to regulate how deep you drive the screw.

The electric screwdriver and the screw gun give you more control than an electric drill with a screwdriver bit. Control is especially important when installing drywall. If you drive a screw too deep, the screw head will tear the facing paper.

With the proper bit installed, both the electric screwdriver and the screw gun will hold the screw for you, leaving your other hand free. Some rely on the fit of the bit in the head of the screw. Others use a magnetic bit.

Cordless electric screwdrivers give you more freedom of movement than the cord type. The disadvantage of cordless models is that the batteries may go dead in the middle of a job. Always carry a spare battery pack. You can recharge a dead battery pack overnight. Some cordless models have a locking collar that allows you to use the tool as a manual screwdriver by turning the entire tool.

Knives

When installing drywall, you'll use knives for scoring and cutting the drywall panels. You'll also use them for applying joint compound or texturing compound.

Cutting knives— You can use just about any type of knife for scoring and cutting drywall, as long as the knife has a sharp blade. The most versatile and practical cutting knives are the standard utility knife and retractable-blade utility knife shown in Figure 2-7.

Each of these knives has a hollow metal handle and a short, heavy-duty metal blade. The blade resembles a single-edged razor blade. The handle is held together with a short metal screw. Inside the handle is a safe and handy place to store

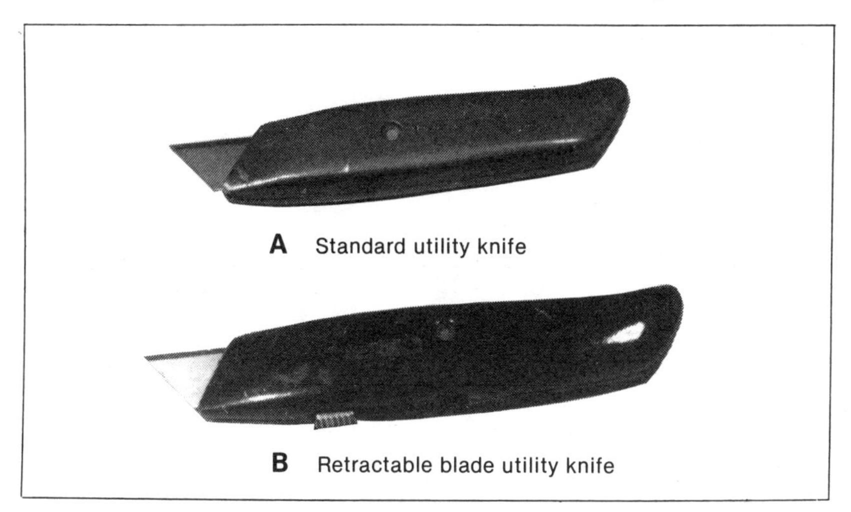

Cutting knives
Figure 2-7

blades. Only half of a blade is exposed at a time, so you can reverse the blade when one end gets dull. The short blade and the large handle that allows a full grip make these tools easy to control.

If you're a drywall hanger who puts the knife in your pants pocket, make sure you're using a retractable-blade utility knife. If you're storing it in one of the pockets of your leather belt pouch, you can use either type.

Stripper— When you need to cut a narrow strip from the edge of a drywall panel, use a stripper. This tool is also known as an edge cutter. It has two serrated wheels and a handle. The two wheels cut the paper on both sides of the drywall at the same time. The handle rides along the edge of the drywall and guides both wheels. Then you just snap off the drywall at the cut.

Drywall knife— This knife is also known as a putty knife, taping knife, or compound knife. You'll use it for applying and smoothing joint compound, topping compound or texturing compound.

This knife is available in widths from 1 inch to 14 inches or more. Figure 2-8 shows 1-inch, 3-inch, 6-inch and 10-inch wide knives. You'll need to have at least three different sizes on hand.

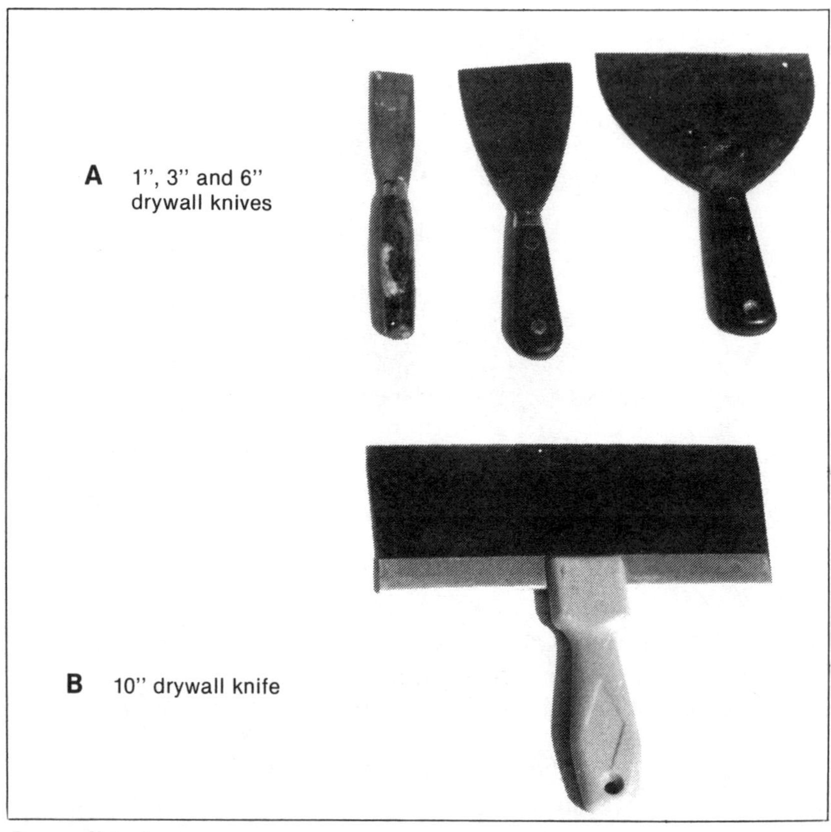

Drywall knives
Figure 2-8

With a little experimenting, you'll find the widths that work best for you. But here are some general guidelines. A 1- or 1½-inch wide knife is handy for patching over nail or screw heads and for filling narrow cracks. A 3-inch wide knife is handy for applying the embedding coat of joint compound and for embedding the tape. A 6-inch wide knife is best for the second coat of joint compound. For feathering out the final coat of joint compound, you'll probably want a knife wider than 6 inches. You may use all of these widths when you're applying texture to finish the drywall.

Drywall knives have handles made of wood or plastic. The handles are riveted or bonded to the blades. The blades are made of spring steel. Blades are available with different degrees of

Trowel
Figure 2-9

stiffness. A very stiff blade will have a different "feel" than a flexible blade when you're working. Select the blade that works best for you.

Trowel— Instead of a wide drywall knife, you may prefer to use a trowel. See Figure 2-9. Some trowels are made especially for drywall applications. They have blades that are slightly springy and concave.

You can use a trowel effectively for feathering the final coat of joint compound and for applying some types of texture. The blade length may be 10 inches or longer. For some applications, you may need both a wide drywall knife and a trowel to complete the job.

Saws

You usually can cut drywall to size by scoring the paper on one side and breaking the board along the score line. This works fine for all your straight-line cuts that run from edge to edge. But you'll also need to make cuts that don't run from edge to edge. And you'll have to cut out openings that don't touch any edges at all. In these cases, use a saw.

Here's how you might use a saw to make a cutout for a wall outlet. First, mark the location of the outlet on the panel. Next, drill a hole at one corner of the section of panel to be removed. Then insert the saw blade in the drilled hole. Cut along the line you marked. This requires a saw blade that's small and easy to maneuver. The saw teeth should be coarse enough to avoid becoming clogged with gypsum as you cut. But they shouldn't be so coarse that they leave the face paper ragged and torn.

Three saws you're likely to use in your drywall work include: a drywall saw, a saber saw, and a circular saw.

Drywall saw— There are hand saws designed specifically for cutting drywall. They have blades that are tapered to a point like a keyhole saw, but the drywall saw handle looks more like one you'd find attached to a file. The handle is a formed cylinder of wood or plastic. It's a straight extension of the blade. This allows you to maintain a symmetrical grip on the saw, even when turning corners. The keyhole saw shown in Figure 2-10 is also handy. You can use it in place of the drywall saw.

Saber saw— One of the handiest saws for drywall work is the saber saw, also shown in Figure 2-10. This is an electric saw with a reciprocating blade that's about 2½ inches long and 5/16 inch wide. Saber saw blades are replaceable and come with different types of teeth. Use closely spaced teeth for cutting metal. Use coarse or widely spaced teeth for softer materials. The closer the teeth, the finer the cut. Blades with medium-spaced teeth work best for cutting gypsum products.

You can also use the saber saw to make cutouts for outlets and switches that don't touch on panel edges. And the advantage to using the saber saw is that you don't need to drill starter holes first. Just turn the saw on and move the edge of the blade through the drywall. Using the saber saw for cutouts can save a lot of time.

The disadvantage of the saber saw is that you need an electrical power source. On some job sites, this may not be possible. But as long as you have power, the saber saw is a good tool to have on hand.

Circular saw— This saw is useful for long cuts that are very close to the edge of the drywall, when there's not enough material left to give you leverage to break the drywall. You can also use it to remove a section of drywall from an existing wall. A circular saw set to the exact thickness of the drywall to be cut saves times and makes a clean, precise cut. Figure 2-10 also shows a circular saw.

Saws
Figure 2-10

The disadvantage of the circular saw is that it throws a huge amount of gypsum dust, creating unhealthy air in the work area. When using a circular saw, be sure everyone in the area wears a breathing mask.

As the name implies, this saw has a circular blade, usually 7¼ inches in diameter. There's a platform on the saw so you can adjust for a specific angle and depth of cut. Rest this platform on the surface of the material to be cut. A fixed guard covers the upper half of the blade. A movable guard covers the lower half of the blade. As the saw moves through the work, the spring-loaded lower guard moves up inside the upper fixed housing. Here's how to use a circular saw to start a "plunge" cut. With the back edge of the saw in a raised position, place the front edge of the platform against the drywall panel. Hold the lower blade guard out of the way in the retracted position. Start the saw and gently lower the blade into the work. Once the blade has fully penetrated the panel, the lower blade guard will be held out of the way by the panel you're cutting.

SPECIALIZED TOOLS

There are a number of tools made especially for drywall work. Most of these tools are expensive, so you might want to consider renting them. Rental rates may be less than what your monthly payments would be if you bought the tools. Consider also the cost of replacing parts that get the most wear.

Many of the specialized tools have either long handles or extension handles that allow you to use them from the floor. That's easier than moving a ladder around and can save many labor hours.

There are seven types of specialized drywall tools: mechanical taping tools, corner joint finishing tools, mixing tools, shaping tools, caulking guns, stud finders and lifting devices.

Mechanical Taping Tools

If you're applying a lot of tape, use a mechanical taping tool. Some of these tools dispense tape and joint compound at the same time. Others dispense only tape. Let's look at the mechanical taping tool, joint tape banjo and dry taping tool.

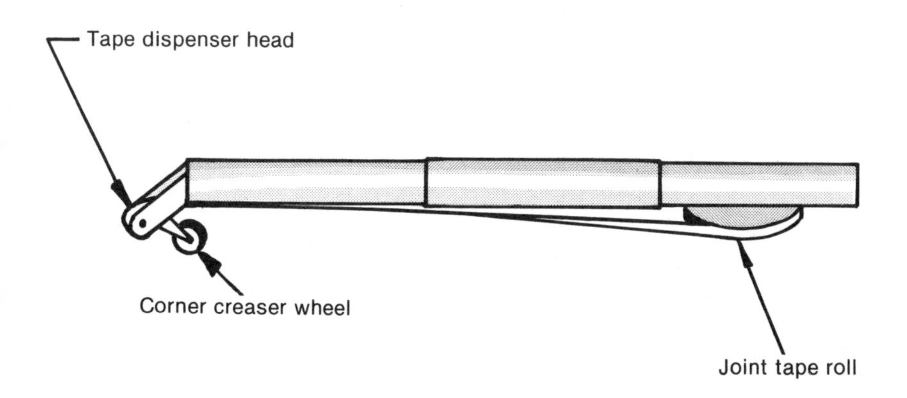

Mechanical taping tool
Figure 2-11

Mechanical taping tool— Figure 2-11 shows a standard mechanical taping tool. This type of tool dispenses joint compound and tape at the same time. The tape is coated with joint compound as tape is applied. When you're using a mechanical taping tool, the joint compound must have a thinner consistency than you'd use for applying joint compound by hand.

There's a knife blade attached to the dispensing end of the tool. Use this blade to cut the tape when you reach the end of the joint. The dispensing end of the tool also has a creasing wheel you can use for taping the inside corners of ceilings and walls. Once you've applied the tape using a mechanical taping tool, use a wide-blade taping knife to smooth the joint.

The easiest way to load joint compound into the tool is with a hand pump designed for joint compound. Using this pump will save you a lot of time and mess.

Joint tape banjo— A banjo is another hand-held tool you can use to apply joint tape and joint compound at the same time. It's simpler than the standard mechanical taping tool. There are two types of banjos, a wet banjo and a dry banjo. Figure 2-12 shows a wet joint tape banjo.

The wet banjo holds the roll of tape and the joint compound in the same compartment. The dry banjo holds them in separate compartments. In either case, joint compound is applied to the

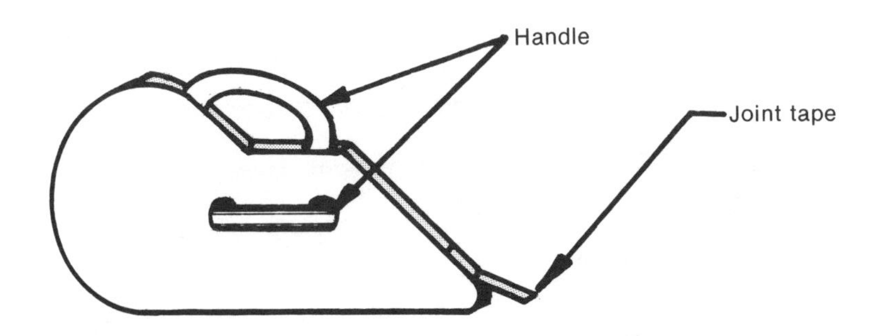

Joint tape banjo
Figure 2-12

tape as tape is withdrawn from the tool. Both banjos have blades to cut the tape at the end of the joint. And they both apply tape and joint compound faster than you can apply them separately.

Dry taping tool— There's a dry taping tool you can use to apply just the joint tape without compound. This tool has a tape-cutting blade to help speed the work. There's also a tool designed just for the application of joint and taping compound to a flat joint. It doesn't apply the tape for you. But it can still speed up your work considerably, even if you apply the tape by hand.

Corner Joint Finishing Tools

When finishing a corner joint with a hand-held taping knife, it takes good control to finish the joint smoothly. Special corner finishing tools can make the job easier, faster, and give a top-quality result. Two kinds are shown in Figure 2-13.

Inside corner tool— This tool is designed for applying joint compound and finishing compound to inside corners. Even if you apply tape by hand, this applicator will speed up your work. The tool for finishing inside corners is bent at slightly more than a 90-degree angle. Because the blade is made of

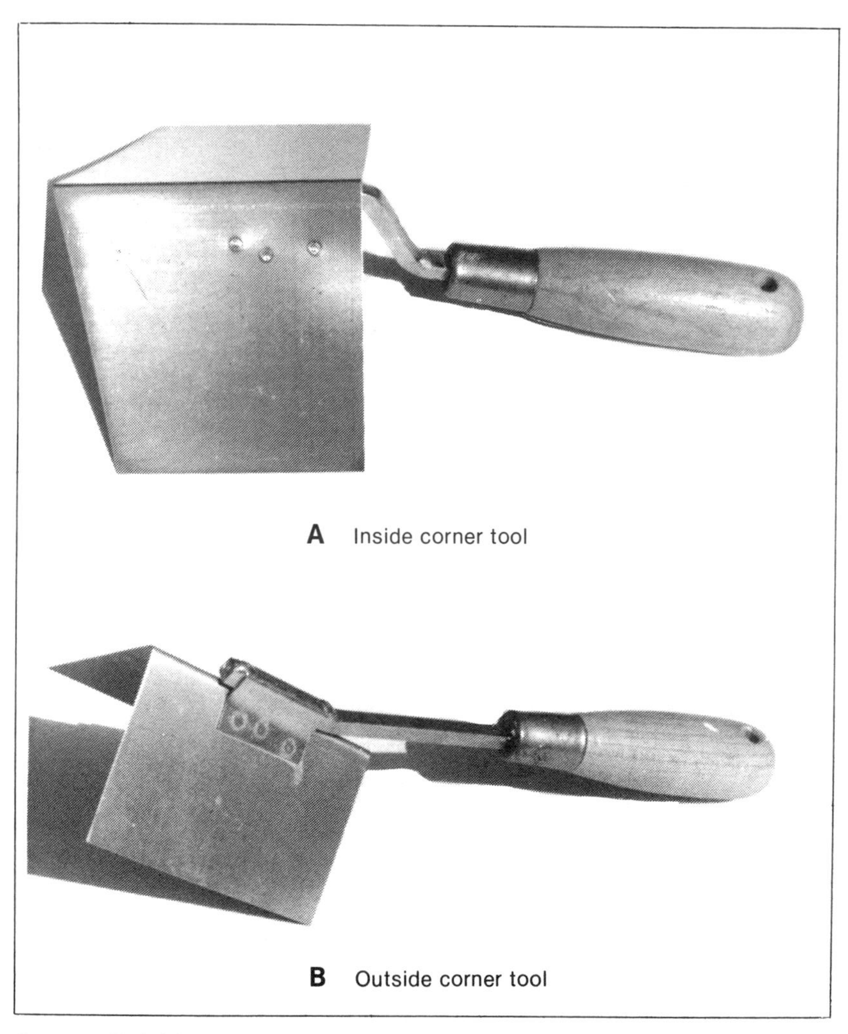

A Inside corner tool

B Outside corner tool

Corner finishing tools
Figure 2-13

spring steel and the angle is slightly greater than 90 degrees, the blade halves will press firmly against both sides of an inside corner. This helps you spread the joint compound evenly.

Outside corner tool— The tool for outside corners has a blade that is bent at slightly less than 90 degrees. This allows the blade halves to press firmly against both sides of an outside corner.

Mixing Tools

Special mixing tools have been designed to fit into a power drill. These tools speed up the mixing of dry mix joint compound, topping, and texturing compounds. They also make it easier to mix compound thoroughly.

Shaping Tools

Two common types of shaping tools you'll use in your drywall work are rasping tools and sanding tools.

Rasping tools— You'll use coarse files, known as *rasps*, for shaping and smoothing the cut edges of drywall. Figure 2-14 shows a standard rasp and a plane-shaped rasp. These tools are effective for many smoothing and shaping operations.

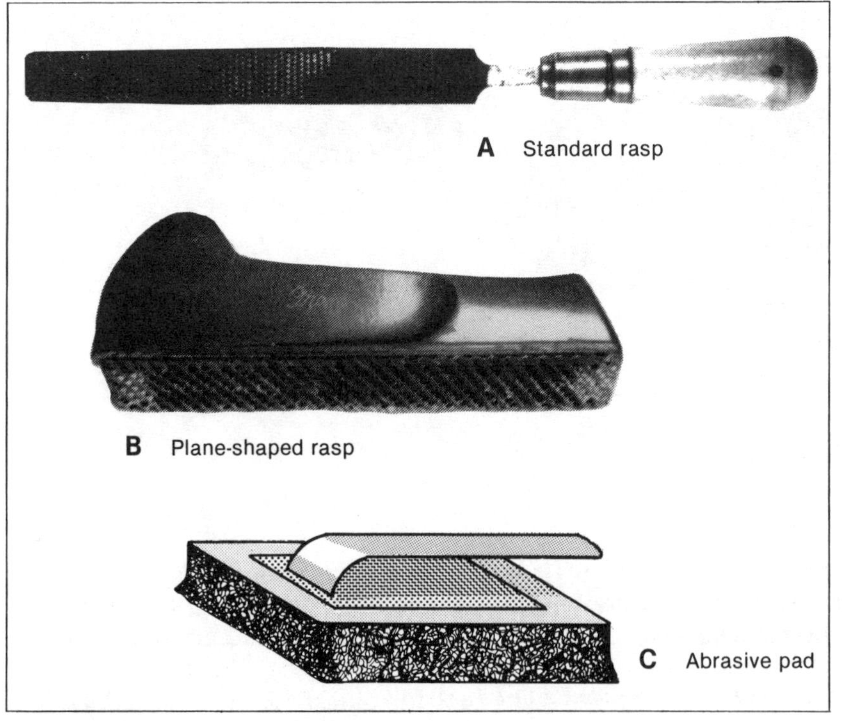

Shaping tools
Figure 2-14

Sanding tools— No matter how skilled you are at applying joint compound, you'll still need to do some sanding. A standard sanding block can do the job for you. Using a block is better than trying to hold loose sandpaper in your hand. You can buy sanding blocks attached to long handles. With this type of block, you can reach just about anywhere on a wall or ceiling.

You can also use abrasive pads made of a woven plastic material. They come in varying degrees of coarseness. These pads are available in sheets or in a rectangular section that has a handle attached. See Figure 2-14. This plastic material is good for sanding and smoothing joint compound. With standard joint or topping compound, you can use the woven plastic pads to *wet sand* effectively.

To do this, dip the pad in water and shake off the excess. Sand the compound much as you would with sandpaper. Moist compound is soft and pliable and smooths more easily. Of course, the compound will have to dry again, but it won't take as long as when you first applied it. The advantage of wet sanding is that it doesn't raise any dust. The disadvantage, of course, is that it doesn't work on water-resistant joint compound.

Caulking Guns

A caulking gun is a handy tool for applying sealants and adhesive compounds. Just insert a cartridge full of compound into the gun. When you squeeze the trigger, it forces compound out the nozzle end of the cartridge. You can cut the plastic nozzle end of the cartridge to different sizes and shapes, depending on what you're caulking.

Stud Finders

There are five common ways to locate studs or joists behind a wall that is already installed. The time-honored methods of tapping on the wall surface and listening for hollow spaces or drilling test holes with a small drill can be both inaccurate and messy. Use a *stud finder* to locate hidden studs or joists. It helps you locate structural members quickly and easily. There are three types of stud finders: magnetic, electronic and ultrasonic.

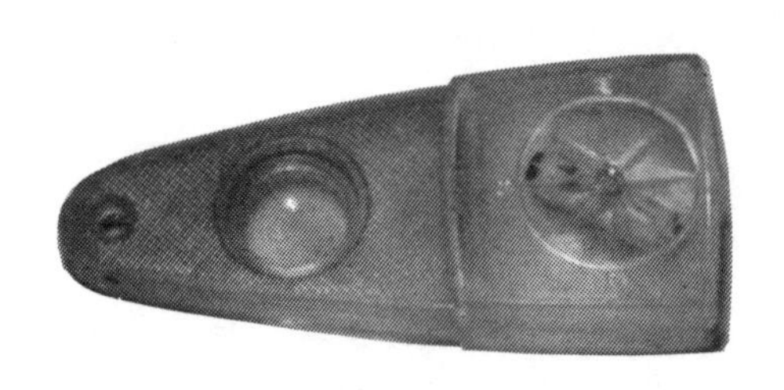

A Magnetic stud finder

B Electronic stud finder

Stud finders
Figure 2-15

Magnetic stud finder— This is the simplest (and probably least accurate) type of stud finder. It consists of a small magnet mounted on a base and supported in the middle like a compass, so it's free to rotate. See Figure 2-15. Move the magnetic stud finder along the wall surface. When you come to a nail head, the magnet will be attracted to it, causing one end of the magnet to swing toward the nail. Since nails attach drywall to the stud, the nail should be near the stud center.

To locate the remaining studs, you can continue moving the stud finder along the wall surface. Or, if you know how far apart the studs are spaced, you can just measure from the first stud. If you decide to measure from the first stud, here are two important points to remember. Most studs are spaced 16 inches on center, but some applications require 24-inch spacing. Studs may be closer together near an intersection with another wall, near a door or a window.

Electronic stud finder— This type of stud finder is slightly more expensive and more accurate. It has a light or a beeper or both. It's also shown in Figure 2-15. The light or beeper is activated when you come across a nail head or other steel. The electronic stud finder is easier to use than the magnetic stud finder and you can adjust the sensitivity.

Ultrasonic stud finder— This type of stud finder is the most expensive and the most accurate. You'll pay about $20 to $30 for one. It locates studs and joists by means of ultrasonic waves. It senses the change in density in the wall when you come across a structural member. This allows you to find the exact location of the stud or joist and determine its exact width.

Lifting Devices

Lifting a ceiling panel into place is no easy task. Drywall panels are heavy and almost impossible for one person to lift into place alone. There are lifting tools specifically designed for placing drywall panels in ceilings.

Figure 2-16 shows a panel hoist that holds drywall and raises it into place. It's on casters, so you can roll drywall to the exact location before installing the fasteners. This device allows one person to install ceiling panels. It's also a great help even when there's more than one person available to do the job.

Safety Devices

It's important that you protect both yourself and your crew when working with drywall. You'll be exposed to gypsum dust particles in the air. And when working with tools, there's always the risk of a loose fragment finding its way into an eye. Always wear eye protection on the job. Use safety glasses with side

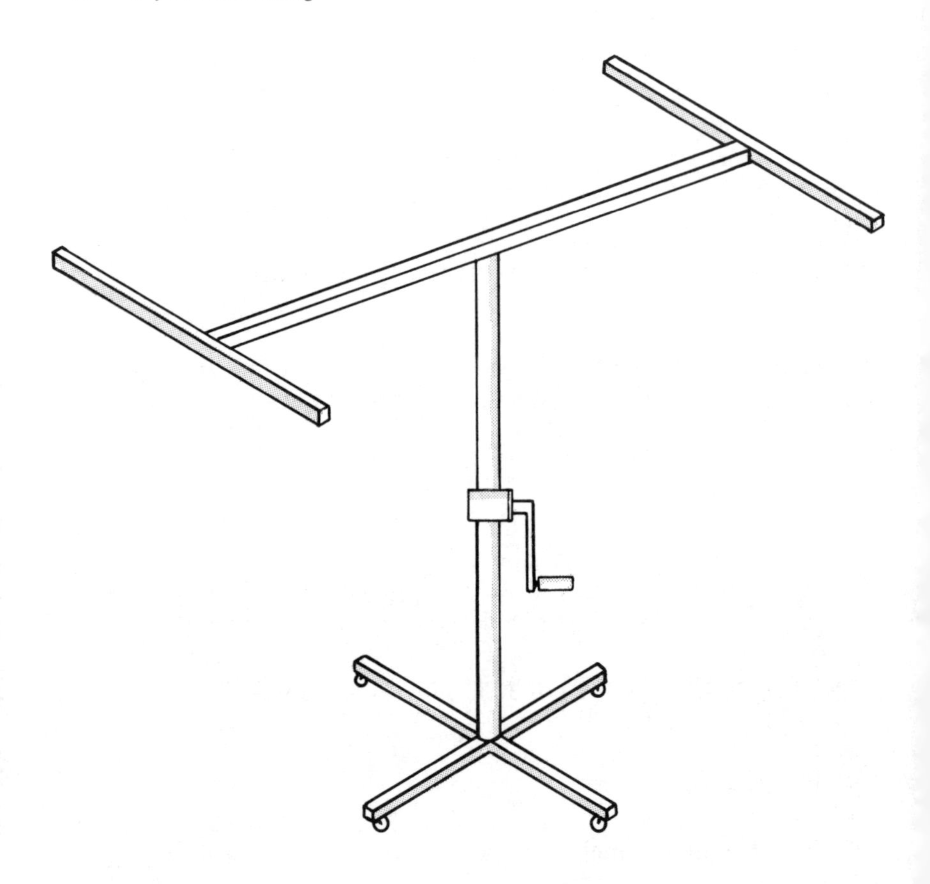

Panel hoist
Figure 2-16

shields or wrap-around safety goggles. When sanding or cutting drywall, everyone in the area should wear dust masks. They're cheap protection. Buy a good supply and use them regularly.

Chapter 3

INSTALLATION METHODS

In this chapter we'll get down to business, applying single-ply and multi-ply drywall. We'll begin with a look at the way drywall should be handled and stored. Then I'll explain what you need to know about the structural base drywall panels are applied to. We'll learn how to measure, mark, cut and fasten drywall panels. I'll suggest installation guidelines and walk you through a typical job. Finally, I'll discuss five specialized drywall applications.

HANDLING AND STORING DRYWALL MATERIALS

I usually use a wallboard dolly to move drywall panels. It's easier and safer than carrying panels by hand. With a dolly, one person can move several at the same time. Figure 3-1 shows one type of drywall dolly.

Move all the panels for one room into the center of that room and stack them flat on the floor. This reduces the amount of

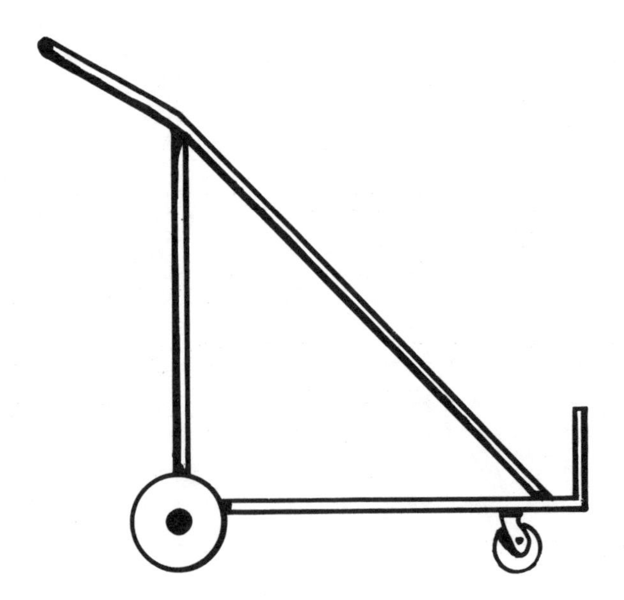

A type of drywall dolly
Figure 3-1

handling required and minimizes damage. Normally you'll stack the ceiling panels on top of the other panels. But don't stack long panels on top of shorter panels. Also, lay them *across* the floor joists to distribute their considerable weight over a greater number of joists.

Always store drywall panels flat. Storing them vertically will damage the bottom edge of the panels. If you want to place risers under stacks of panels, use blocks of wood, usually 2 x 4 stock, at least 4 feet long. Place them under the stacks of drywall panels no more than 28 inches apart to provide spaces for a forklift to insert its tines. Put the risers across the width of the panels, extending from one edge to the other, as shown in Figure 3-2.

Try not to store non-water-resistant gypsum panels outdoors. Exposure to the elements can cause them to deteriorate. If you do have to store panels outside, provide complete protection against the weather. Choose a storage area that's firm and has good water run-off. Place a row of risers on the surface of the storage area under the first layer of gypsum panels. Cover the stacks of panels with watertight material, but leave some space

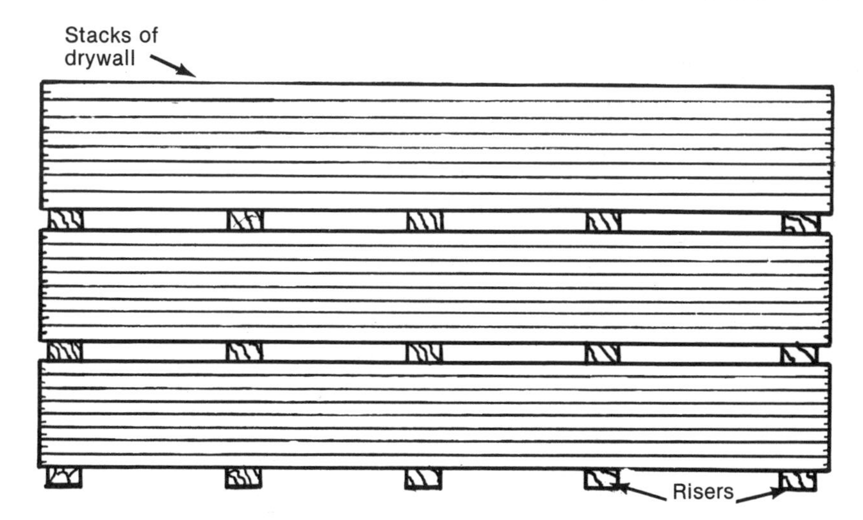

Stacking drywall with risers
Figure 3-2

open near the ground for air circulation. Tie the covering in place to make it secure. These precautions may not completely prevent damage to the gypsum panels, but they'll minimize it.

Don't have the panels delivered to the job until you're ready to begin work. Panels delivered too early are often damaged before installation begins. Order other materials, such as compounds and veneer plasters, so they arrive just before you're ready to use them. These materials deteriorate with age. Use the freshest materials possible. Store beading where it won't get bent. Store all materials, especially compounds, out of the sun and weather where they won't be exposed to extreme changes in temperature.

SINGLE-PLY CONSTRUCTION

Single-ply drywall uses just one layer of gypsum wallboard applied to ceilings or walls. This is the most common type of wallboard installation in residential and commercial buildings. It covers the framing but provides only limited sound insulation, fire resistance and water resistance.

Base Materials

You can install drywall over just about any base as long as the base is flat and stable. That's important. A flat, stable base is needed to keep the finished drywall surface smooth and flat.

Even though a drywall panel adds strength to a wall, the base or framing has to be strong and firm before you begin work. The base must be strong enough so you can attach the drywall to it. And the base has to be stable enough so that it doesn't shift later, causing cracks in the drywall or joints.

Three common base materials are concrete, wood framing and metal framing. Let's look at each of these.

Concrete and masonry— If you're applying drywall directly to concrete or masonry, make sure the surface is smooth and flat. If it's not smooth, apply furring strips to the face of the concrete or masonry so there's a flat, even surface for the drywall. If you think that moisture may come through the concrete in wet weather, apply a moisture barrier and furring strips before hanging the drywall. Use asphalt felt or kraft paper as the moisture barrier. Then install the furring strips vertically on the masonry wall. Be sure to install the drywall panels perpendicular to the furring strips.

The best way to attach furring to masonry is with power-driven concrete nails. It's fast and cost-effective. But you can also drill holes in the concrete and use a number of different anchor inserts and screws to hold the furring to the wall. Either way, space the fasteners 16 inches on center for 1-inch stock. For 2-inch stock, you can space them up to 24 inches on center.

Use at least 1 x 2-inch furring strips over masonry and concrete. This size is thick enough for the drywall fasteners to grip. It also raises the drywall far enough from the masonry so that the moisture can't reach the drywall.

The normal curing time for concrete is 28 days. Wait until the concrete is at least this old before you apply drywall to it.

If the masonry wall is above grade, dry and smooth, you don't need furring strips. Use adhesives to laminate the drywall directly to the masonry wall. The exception is when you're applying decorated wallboard panels with a watertight covering such as plastic. Here, moisture might become trapped within the core of the drywall, causing it to deteriorate. So always use furring strips with the decorated panels.

Wallboard thickness	Installation method	Framing member spacing (o.c.)
Ceilings:		
3/8"	Perpendicular	16"
1/2"	Perpendicular	16"
1/2"	Parallel	16"
5/8"	Perpendicular	24"
5/8"	Parallel	16"
Walls:		
3/8"	Perpendicular	16"
3/8"	Parallel	16"
1/2"	Perpendicular	24"
1/2"	Parallel	24"
5/8"	Perpendicular	24"
5/8"	Parallel	24"

Wallboard thickness and maximum spacing of framing members
Figure 3-3

Wood or metal framing— On most jobs you'll apply drywall to wood or metal studs or joists. In some cases, however, you'll install self-supporting drywall partitions. These are partitions that don't use framing members. Instead, the outer gypsum panels are fastened to 1-inch thick gypsum coreboard panels held in position by metal runners attached to the floor and ceiling of the structure.

Whatever the framing material, it must be properly spaced. The farther apart the framing members are, the thicker the drywall must be. Figure 3-3 shows the relationship between drywall thickness and the maximum allowable spacing for framing members. Notice that the installation method may affect the spacing of the framing members. When you install 5/8-inch thick ceiling panels perpendicular to the framing members, you can increase the spacing between members to 24 inches. If you install the same drywall parallel to the framing members, maximum spacing is 16 inches.

All framing members should be aligned in the same plane. This means that when you sight across the framing members, they should be in a straight line. No stud or joist should be more than 1/4 inch out of alignment with other studs or joists in the wall or ceiling. This is important. Drywall can't compensate for framing that's out of alignment. Misaligned framing members will show up in your finished wall surface.

If your framing members are out of alignment, do some repair work before you begin installing the drywall. If only a few studs or joists are out of alignment, replace them. You can also place small strips or shims to even up the face of the wall. But if a stud or joist sits too far in on one side of the wall, there's a good chance it's too far out on the other side of the wall. Look at both sides of the wall before you do any shimming. It may be easier to replace the stud or joist.

You may be able to straighten a warped stud by making a saw cut on the concave, or hollow, side of it. Then push the stud into line and drive a wedge into the saw cut. Reinforce the stud by nailing wood reinforcing strips on both sides of it. The wood reinforcement should be at least 1½ feet long and at least 3/4 inch thick. A 1 x 4 makes good reinforcing strip.

You may not have to replace a misaligned ceiling joist. Instead, bring it back into alignment. Lay a straight 2 x 6 on edge across the top of the ceiling joists. Then toenail the 2 x 6 to each of the joists. This will usually bring the misaligned joist back into line with the others.

If more than one framing member is out of alignment, your best bet may be to install furring strips over the framing. You can shim the furring to make a flat surface. The spacing for furring strips can't be any greater than the spacing for framing members.

If the furring is fully supported along its length, it will provide a sturdy surface for nailing. Furring applied along the length of the studs is called *parallel furring.* Its purpose is usually to increase the thickness of the wall. You can also apply it selectively to some studs to correct alignment problems.

Furring applied *across* studs or other framing is known as *cross-furring*. This type of furring will *rebound* when you drive nails into unsupported furring sections. Furring strips smaller than 2 x 2 have too much rebound for nail fasteners. In this case, use screws to fasten drywall to the furring. If you use screws as fasteners, you can use smaller cross-furring members than when you use nails.

Fastener type	Minimum grip length
Smooth-shank nails	⅞"
Ring-shank nails	¾"
Screws (wood)	⅝"
Screws (sheet metal)	⅜"
Screws (gypsum)	¾"
Staples	⅝"

Minimum grip length for drywall nails and screws
Figure 3-4

The furring must be thick enough to provide the required fastener *grip length*. This is the length of the part of the fastener that's buried in the framing. Figure 3-4 shows minimum grip lengths for nails, screws and staples used to install gypsum drywall. Keep in mind that the screw point isn't included in the grip length. Figure 3-5 shows the required fastener lengths for various drywall thicknesses.

When using wood framing as a base for drywall, be sure the framing members are the proper size and grade. Use 2 x 4 (or larger) studs in all load-bearing walls to ensure that the walls have adequate strength. For non-load-bearing walls, you may be able to get by with 2 x 3 studs. The wood should have a low moisture content, usually 15% or less. If the moisture content is higher than 15%, the studs may shrink and warp as they dry, ruining your drywall job. The wall surface may be uneven or cracks may develop. Shrinkage will also loosen nails.

In areas where fixtures or heavy objects may be hanging from the drywall, be sure to add structural support to the framing before installing the drywall panels. For example, provide extra support behind a grab bar over a tub, or where a sink will be hung on a wall.

To provide the extra support, install additional framing members between the studs. At the point where a sink will need support, fasten 2 x 8 or larger stock between the two studs on either side of the sink location. Orient the new member so that its widest dimension (the 8-inch width) is vertical and flush with the face of the studs. This additional support will transfer the

Fastener type	Wallboard thickness	Minimum fastener length
Smooth-shank nails	3/8"	1-1/4"
	1/2"	1-3/8"
	5/8"	1-1/2"
Ring-shank nails	3/8"	1-1/8"
	1/2"	1-1/4"
	5/8"	1-3/8"
Screws into wood	3/8"	1"
	1/2"	1-1/8"
	5/8"	1-1/4"
Screws into sheet metal	3/8"	3/4"
	1/2"	7/8"
	5/8"	1"
Screws into gypsum partitions	3/8"	1-1/8"
	1/2"	1-1/4"
	5/8"	1-3/8"
Staples into wood	3/8"	1"
	1/2"	1-1/8"
	5/8"	1-1/4"

Minimum fastener length for various wallboard thicknesses
Figure 3-5

weight of the sink to the studs, and to the rest of the structure. Without it, the drywall itself wouldn't be strong enough to support the weight.

Once the framing is installed, wait as long as you can before installing drywall. This gives the lumber a chance to give off excess moisture before drywall is hung. Lumber dries quickest when the temperature is 55 degrees or higher. Warm weather will also help the framing stabilize. After the drywall is up, provide air circulation so the framing and drywall compounds can dry evenly.

Before installing the drywall, check to be sure all service lines in the wall are placed properly. Plumbing, electrical and HVAC ducting can't protrude beyond the face of the framing. Where service lines pass through the framing members, it's a good idea to install metal cover plates on the studs. These plates come in a variety of lengths. They have pointed cutouts on the plate, bent out at 90 degrees from the plate surface (Figure 3-6). This lets you pound the plates into wood studs without using nails. The plates will protect service lines from fastener damage when you're installing the drywall panels.

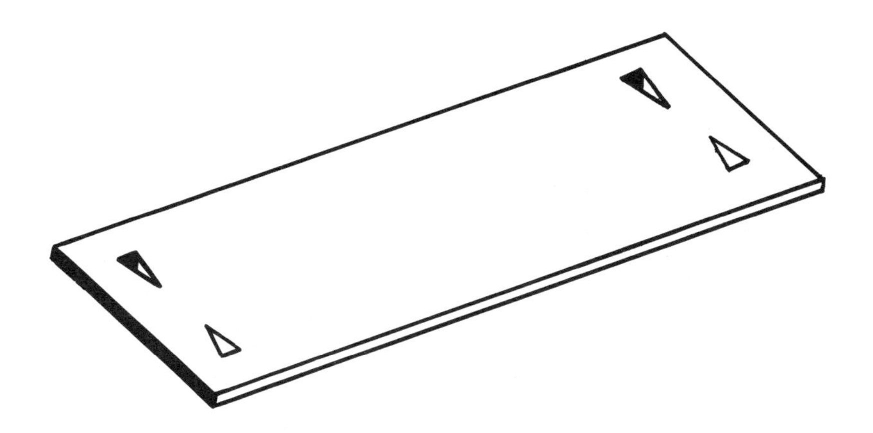

Metal cover plate
Figure 3-6

Measuring and Marking the Drywall

In large rooms and on large projects, you'll use many whole sheets of drywall. But you'll also have some cutting to do, regardless of the size of the job. The first step in making accurate cuts is to make accurate measurements. Measure carefully the area to be covered and then transfer these measurements to the panel. Sounds pretty easy, doesn't it? Unfortunately, it's not, as any experienced drywall installer will admit. But inaccurate measuring and marking have ruined thousands of sheets of perfectly good drywall. The cutting itself is seldom the problem. There's real wisdom to the old saying, "Measure twice and cut once."

There are three common reasons for cutting sheets of drywall. You'll need panels that are shorter or narrower than the standard size. You'll need to cut openings for outlet boxes. And you'll need to make angle cuts to fit the drywall to angled surfaces. Before we look at how to measure and mark the drywall for these three types of cuts, let's review the tools commonly used for marking drywall. The three tools you'll use most often are pencil, pen and a chalk line.

Pencils are O.K. And pens work fine too, as long as they have waterproof ink. If the ink is water soluble, the joint compound will smear it and the line will show through when the job is finished. The trouble with permanent ink is that it's often harder to cover up than pencil or chalk.

When using a chalk line, be careful about the color chalk you use. Of the several colors available, red is the hardest to cover. Blue is probably best.

Here's how to use the chalk line. First measure and mark two points on the panel you're going to cut. Make the marks near the edges of the panel. Attach the chalk line hook to one edge of the drywall so that the string will lay across the first mark you made. Lay the chalk line across the panel so that the string also lays across the second mark and spans between the two marks. Pull the string tight enough to take out the slack. Using your other hand, lift the center of the string straight away from the surface of the panel. Let it snap back against the panel, as shown in Figure 3-7. When the string snaps back, it leaves a line of chalk dust across the surface of the panel. This is your cut line.

Cutting smaller panels, cutting openings for outlet boxes, and making angle cuts, present a different set of problems.

Shorter or narrower panels— A tape measure is the handiest measuring device for most situations. Let's say you're covering a rectangular area that's either less than full panel height or less than full panel width. Measure the first dimension on the wall or ceiling you want to cover. Transfer that measurement to the *face* of the drywall panel, near one edge. Make a pencil mark at this point. Move the tape measure to the opposite edge of the drywall and measure the same distance, making another pencil mark. Snap a chalk line between the two marks. That's your cut line.

Snapping a chalk line
Figure 3-7

Outlet box cutouts— First measure the horizontal distance from the near edge of the outlet box to the edge of the drywall panel. Mark this point on the face of the drywall. It's point 1 in Figure 3-8. Measure the width of the box. Add that width to the first measurement and make a second pencil mark on the drywall (point 2). Measure the vertical distance from the near edge (top or bottom, whichever is closer to the edge of the panel) of the outlet box to the edge of the drywall panel. Mark this point on the face of the drywall in line with the horizontal

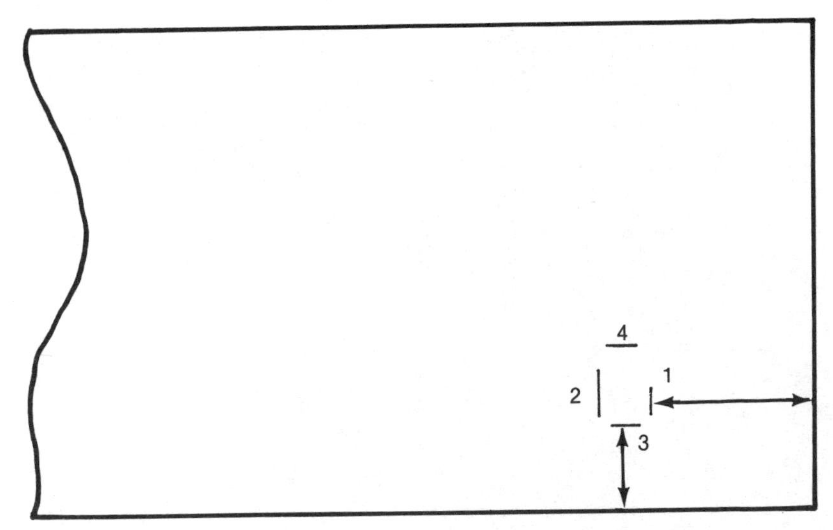

Sequence of measurements for outlet box cutout
Figure 3-8

marks you made (point 3). Measure the height of the box and add this to your other vertical dimension (point 4). Be sure you add it in the proper direction. If you measured the first vertical dimension from the bottom of the panel to the bottom of the box, go up with the height measurement. Otherwise, go down from the first vertical measurement.

You now have marks on the face of the drywall that define the dimensions of the outlet box at the proper location. Extend the horizontal and vertical marks as necessary to make them connect up in the shape of the outlet box.

Angle cuts— When you need to fit drywall to an angled surface, there are two ways to measure for the cut. You can measure down from two points on the angle, or you can use a level protractor. We covered both procedures in Chapter 2. Either way, use a chalk line to mark the cut line on the panel.

If you use a protractor, determine the angle and measure the vertical distance at the location where the edge of the new panel will go. Mark this vertical distance on the face of the new panel. Use the level protractor to mark the angle of the cut on the face of the new panel. Then draw a line following this angle for a distance of at least a foot. Hold one end of the chalk line at the vertical mark on the edge of the panel. Move the other end of

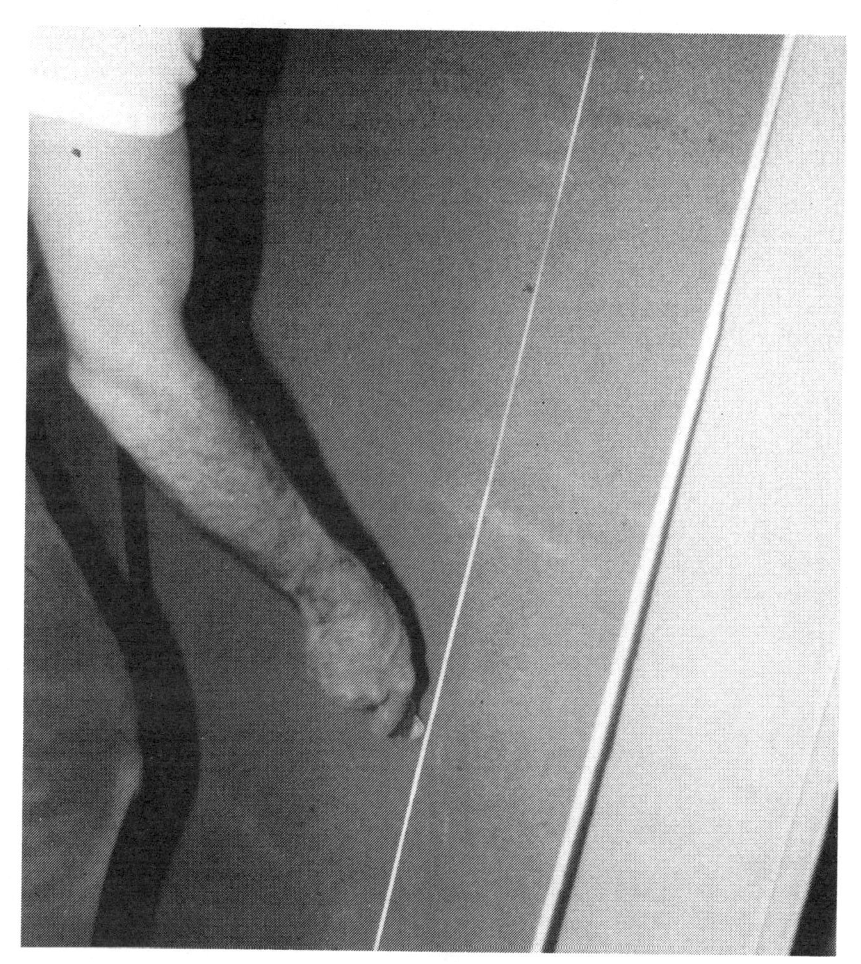

Scoring wallboard with a utility knife
Figure 3-9

the chalk line so it follows the same line you drew with the straightedge. This ensures that the chalk line will be at the correct angle. The chalk line allows you to extend the straight line of the cut quickly and accurately.

Cutting the Drywall

How you cut a drywall panel depends on the type of cut you need to make. The four types of cuts you'll make most often include: a straight-line cut from edge to edge, an edge-to-edge cut that doesn't run in a straight line, narrow strips, and cutouts. You'll also need to know how to repair damaged panels.

Straight-line cut from edge to edge— Use a utility knife to score the facing paper along the cutting line you have marked. See Figure 3-9. You may want to lay a straightedge along the mark to serve as a guide for the knife. Or there are cutting guides that you can attach directly to your tape measure. One guide clamps to the tape and rests on the edge of the drywall panel. Another guide fits on the end of the tape and has a slot for the utility knife blade (Figure 3-10). By setting the tape measure and edge guide at the proper distance for the cut, you can move the tape and the knife along the length of the panel and make a cut in the proper place.

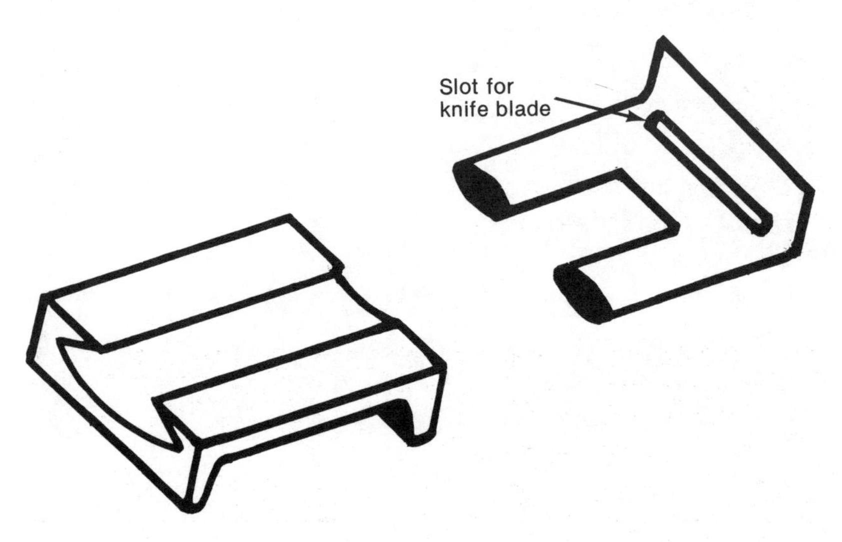

Cutting guide for tape measure
Figure 3-10

Once the cut is completed, using your hands or your hands and knee, apply pressure to the back side of the cut. See Figure 3-11. The panel will snap cleanly along the line you cut in the facing paper. Continue bending the drywall back to slightly crease the backing paper. Cut along this crease. Use a rasp to smooth the cut edge, as shown in Figure 3-12.

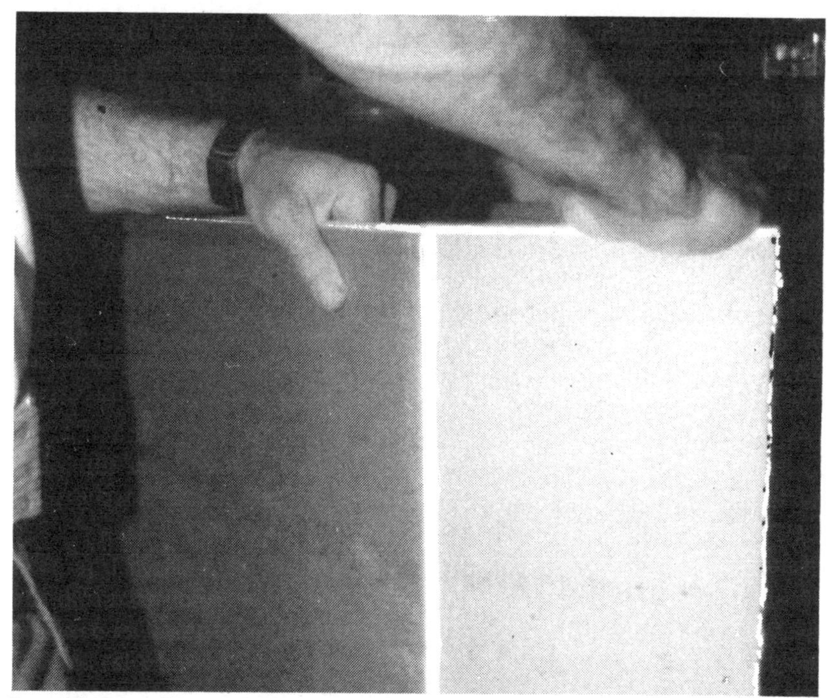

Breaking wallboard along cut line
Figure 3-11

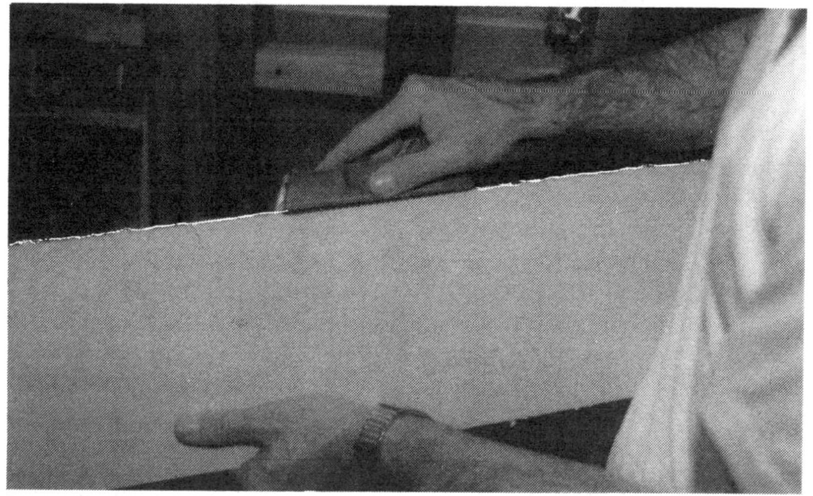

Smoothing cut edge of wallboard
Figure 3-12

Non-straight-line cut from edge to edge— If you have to cut out a section that runs edge to edge but doesn't run in a straight line from one edge to the other, use a saw to make one of the cuts. Use a utility knife to make the other cut and snap the drywall as in the first example. Then cut the backing paper for the second cut.

Narrow strips— When removing a narrow strip along the edge of a panel, you can score it with a knife if the strip is 3 inches or wider. Then snap the drywall in the usual way. If the strip is narrower than 3 inches, it may be hard to snap it cleanly along the cut line. In this case, use a saw that will cut the panel all the way through.

A second way is to measure carefully and score both the facing and the backing paper with a knife before attempting to snap the drywall. This helps make a clean break. Or you can use a tool called an edge stripper to cut away a narrow section of panel. This tool has two cutter wheels that score the front and back of the panel at the same time. There's an adjustable guide to set the width of the cut.

Here's a third method. It takes a little more time than the others, but makes a clean edge with little mess. Just use a knife and make repeated passes at the cut line, taking it deeper each time. You can cut all the way through the panel in this manner.

The final way to remove a narrow strip is to cut it after installation. I do this if the panel extends over the edge of a doorway or window. Use a saber saw to cut the panel back to the edge of the doorway. See Figure 3-13.

Cutouts— Use a saw to make cutouts that don't extend to the edge of a panel. Use a keyhole saw or drywall saw after drilling a starter hole at one corner of the cutout. If you use a saber saw, you can make a plunge cut without drilling a starter hole first. Just lower the blade slowly into the panel while the saw is running. You can rest the back of the saw on the drywall panel while you lower the blade to give you more control.

Repairing damaged panels— You may have to replace a section of damaged drywall in the middle of a finished wall or ceiling. When cutting into an existing wall, make your vertical cuts along a line that runs down the center of a stud. Here's how to do it.

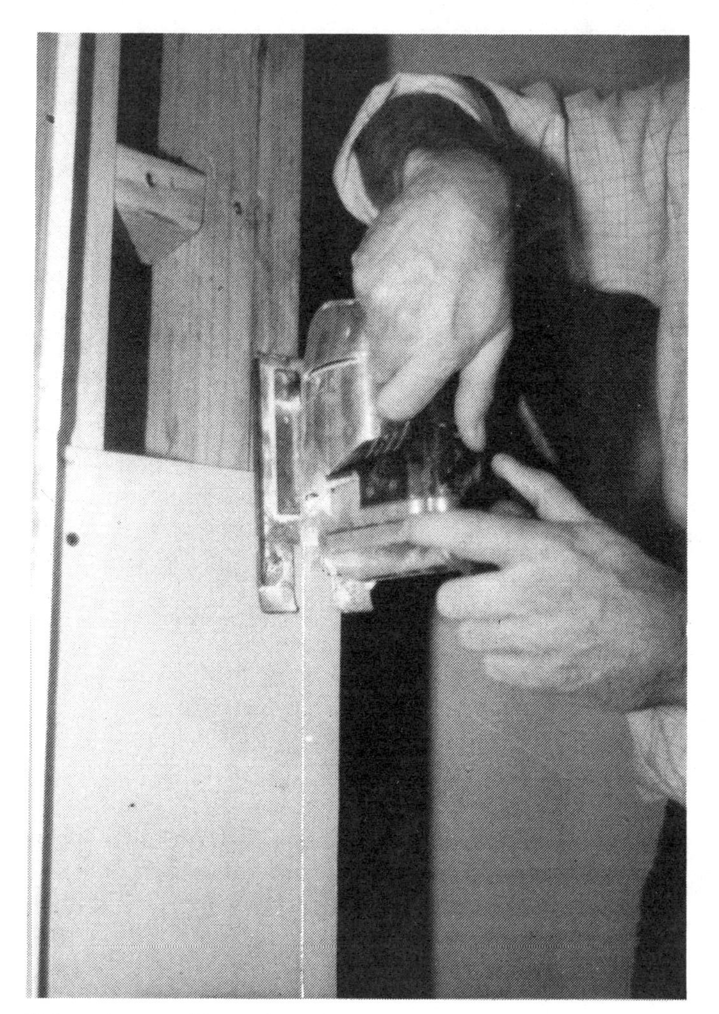

Trimming wallboard at doorway with a saber saw
Figure 3-13

Locate the stud either with a stud finder or by drilling a small diameter hole in the existing panel. You'll be patching the area anyway, so the hole will be covered later. Once you locate the studs, use a level or a plumb bob to mark vertical lines along the centers of the studs. Use a level to connect vertical lines with horizontal lines at the top and bottom of the section you're going to remove.

If you're using a circular saw, set it to the thickness of the drywall before cutting along the lines. This keeps the saw from cutting into the studs. The disadvantage of using a circular saw is that it throws a lot of dust.

You can also use a chisel to cut along the lines. This takes longer but it isn't as messy. You could also use a utility knife, cutting along the line again and again until you cut through the panel. A saber saw is a good tool for cutting a panel along the horizontal lines. But the blade will go deeper than the thickness of the panel. You might cut an electrical wire or water line inside the wall. Don't take the chance. Either use another tool or trim off enough of the saber saw blade so it doesn't penetrate any deeper than the thickness of the panel.

Fastening the Drywall

Drywall can be fastened with nails, screws, staples or adhesives. For many years, nails were the most commonly used drywall fastener. Screws are now common because of their great holding power. It takes fewer screws to hold a panel than it does nails. And you *have* to use screws on steel studs. Staples are a good choice when you're installing the base layer of drywall in multilayer construction. You can use adhesives nearly anywhere nails or screws can be used.

When using nails, screws and staples, correct spacing is important. Proper spacing guarantees that there will be enough fasteners to secure the drywall to the framing. If you don't use enough fasteners, ceilings can sag and walls may bow. Figure 3-14 shows the maximum spacing for nails, screws and staples in gypsum drywall panels.

Fastener type	Application	Maximum spacing
Nails	Ceilings	7"
	Walls	8"
Screws	Ceilings	12"
	Walls	16"
Staples	Ceilings	7"
	Walls	7"

Maximum spacing for wallboard fasteners
Figure 3-14

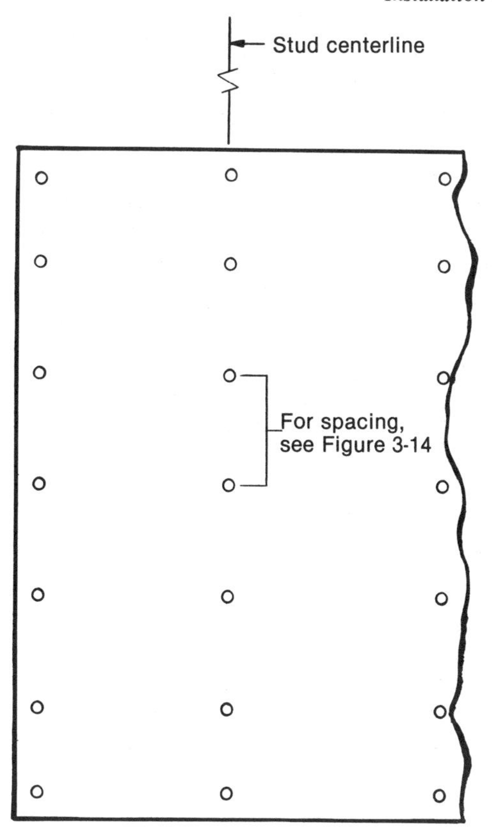

Single nailing
Figure 3-15

Nails— Use nails only when the framing members are wood. Nails are available with smooth shanks or ring shanks. Ring-shank nails have rings forged into the shank of the nail. This increases the holding power of the nail by about 20%.

The length of nail you use depends on the thickness of the drywall. Look again at Figure 3-5. This figure shows the minimum nail lengths for smooth-shank and ring-shank nails. Notice that you can use a shorter ring-shank nail than smooth-shank nail for the same drywall thickness. This is because of the greater holding power of the ring-shank nail.

Use single nailing along the edges of the drywall that are parallel to the framing members. Figure 3-15 shows the single-nailing method of fastening drywall to framing members.

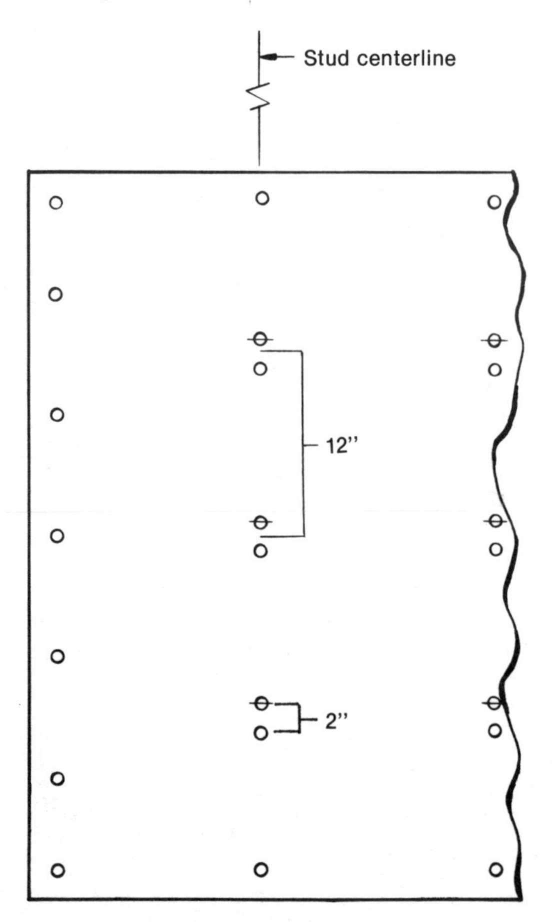

Double nailing
Figure 3-16

Instead of using a single nail at every location, you may want to double-nail some areas for added strength. In this case, you'll drive two nails at one location. Starting at the center of a panel and working toward the edges, first install one nail at each location. After you've single-nailed the entire panel, go back and drive a second set of nails. Space the two nails 2 to 2½ inches apart. See Figure 3-16. Notice that the sets of double nails are spaced 12 inches on center (o.c.).

Screws— Screws can provide up to 350% more holding power than nails. Because of this greater holding power, you can space screws farther apart, using fewer screws to complete the job. Installation is quick and easy with an electric screw gun. Be sure to use screws wherever steel studs are the framing members.

Drywall screws have sharpened points that eliminate the need for predrilled holes. These screws have heads that flare out from the shank. This causes the facing paper to twist slightly under the head as the screw is tightened. Twisting increases the holding power and prevents damage to the facing paper. The screw threads are at a steeper angle than the threads on a regular screw, reducing the number of turns needed to drive each screw. The flat head is easily concealed with joint compound, providing a smooth finished surface.

There are three common types of drywall screws: Type W, Type S and Type G.

- Type W: These screws are specifically designed for fastening drywall to wood framing. They're usually 1¼ inches long and should penetrate at least 5/8 inch into the wood framing.

- Type S: These screws are designed for attaching drywall to sheet metal or metal studs. They have a smaller-diameter shank than Type W screws and come in a variety of lengths. They have a slotted, drill-type point that easily penetrates the metal, eliminating the need for predrilled holes. Type S screws should penetrate at least 3/8 inch into the metal. You can substitute Type S for Type W screws when you need a longer length of Type W screw.

- Type G: Use these screws to attach drywall to gypsum coreboard or a base layer of drywall. These screws have a double thread. This allows the screw to grip the gypsum firmly, resisting pull-out. The screw should penetrate at least 1/2 inch into the base or core layer of drywall.

Staples— You normally use staples only in multilayer construction. They're a practical fastener for attaching the base layer to wood framing members.

The crown of the staple should be at least 7/16 inch wide. This size provides adequate holding power and won't tear through the facing paper on the drywall. The staples should be made of galvanized steel. And the legs should be long enough to penetrate at least 5/8 inch into the wood framing. Use a pneumatic or electric staple gun to drive them into the drywall.

Adhesives— Use adhesives to bond drywall to studs, to a base layer or to a gypsum core. You can use adhesives just about anywhere you can use nails or screws. There are three key advantages to using adhesives:

- Adhesives form a stronger bond than nails. This means you'll have fewer loose panels to repair in the future.
- The adhesive bond isn't altered by changes in temperature or moisture.
- When using adhesive, the installation will require 75% fewer fasteners to hold the panels in place.

The type of adhesive you use depends on the surface you're attaching the drywall to. If you're fastening drywall to framing members, you'll use one type of adhesive. It takes a different adhesive to attach the drywall to another layer of drywall. You can apply the adhesive with a roller, brush or caulking gun.

If you're using a caulking gun, buy the adhesive in cartridges. The cut you make on the end of the cartridge nozzle will be different for wall and ceiling applications. See Figure 3-17.

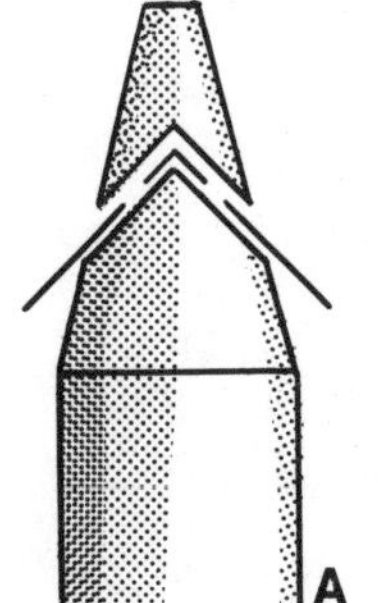

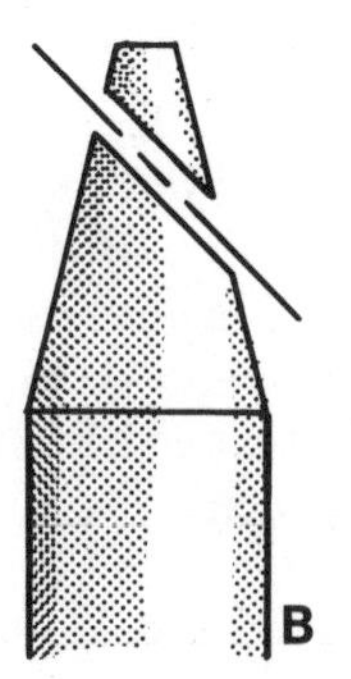

Cutting the adhesive cartridge nozzle
Figure 3-17

For wall applications, cut the end of the cartridge nozzle in an inverted "V," as shown in Figure 3-17 A. This allows you to lay a uniformly round bead of adhesive on the studs. The bead should be about 3/8 inch wide to bond drywall solidly to the framing.

For ceiling applications, cut the end of the nozzle at an angle. See Figure 3-17 B. With the angle cut, you can use a spreading action to apply the adhesive to ceiling joists. This method is best because it keeps adhesive from dripping off the joists.

When attaching drywall to framing, apply the adhesive directly to each stud or joist. Make sure the framing surfaces are clean so the adhesive sticks firmly. There's no need to apply adhesive to diagonal bracing or to top and bottom plates.

All adhesives have an *open time limit*. Once you apply adhesive, it begins to dry, quickly forming a *skin* of partially dried adhesive. It's important to get drywall panels fastened into place before the adhesive begins to set up. Once a skin forms, the sticking power of the adhesive is reduced and the bond won't be as strong. If you don't use all of the adhesive in a cartridge, insert a screw or bolt into the end of the cartridge to seal it.

Always provide adequate ventilation when working with adhesives. Some adhesives use a strong solvent base. Continued breathing of the fumes can be dangerous. Read the cartridge label carefully. And be careful about working near an open flame or sparks. It doesn't take much to start some types of adhesives burning.

You'll need to use fasteners to hold the drywall panels in place until the adhesive sets. Figure 3-18 shows spacing requirements for fasteners used with adhesive. It may not be necessary to use fasteners along the vertical edges if you use prebowed panels. This allows you to install fasteners only at the top and bottom of the panel.

Prebowing drywall panels— When you're installing decorated drywall panels, you won't want to drive fasteners along the vertical joints between panels. But if you limit the fasteners to the top and bottom of the panel, there won't be any fasteners holding the center of the panel in place while the adhesive sets. *Prebowing* the drywall panels will solve this problem.

A properly prebowed panel curves in against the framing at the center of the panel. The top and bottom of the panel won't touch the framing until you install the fasteners. When you hold

Application	Installation method	Fastener spacing
Ceiling	Perpendicular	16" o.c. at each end of panel and at each joist along panel edges. Install one temporary fastener at mid-point of panel at each joist.
Ceiling	Parallel	16" o.c. along panel edges and at each joist at ends of panels. Install temporary fasteners 24" o.c. at joists between panel edges
Wall	Perpendicular	16" o.c. at end of each panel and at each stud along panel edges.
Wall	Parallel	16" o.c. along panel edges and at each stud at ends of panels.

Spacing for fasteners used with adhesive
Figure 3-18

the top and bottom tightly against the framing to install the fasteners, the curve in the panel will hold the center of the panel against the framing.

Here's how to prebow gypsum drywall panels. Stack the panels face up, with the ends resting on 2 x 4 pieces of lumber. Leave the centers of the panels unsupported, as shown in Figure 3-19. Leave the panels in this position until they take a permanent set. Overnight is usually long enough.

If the weather is very humid, the panels may be too flexible to stiffen into a bowed shape. In this case, increase the number of 2 x 4 boards supporting the panel ends. You can also prebow gypsum panels by stacking them face down with 2 x 4 lumber supporting the center of the stack of panels. This lets the panel ends sag.

Decorated drywall panels aren't the only panels that can be prebowed. You may need to install drywall panels on a curved surface such as a curved staircase or curved entryway partition.

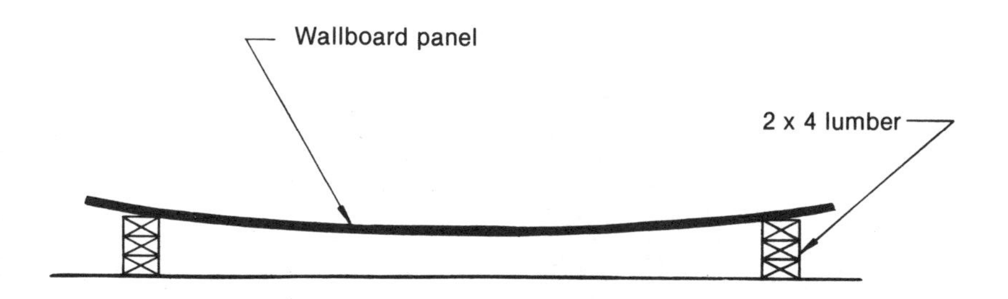

Prebowing gypsum wallboard panels
Figure 3-19

These surfaces usually have a relatively gentle curve to them. For sharper curves, consider installing intermediate studs to prevent flat spots in the drywall between framing members. Prebowing the panels will help you install them tightly against the framing without danger of breaking the panels.

When prebowing panels to fit a curved surface, you can spray the panels lightly with water and let them set overnight. Water will soften the panels and let them bow more quickly. And a damp panel will sag more than a dry one. When the panel dries out again, it'll be just as stiff and strong as before spraying.

Let's look at a sample problem. Assume you're installing drywall in a curved archway. You want to prebow the drywall to cover the top inside surface of the archway. Here are the four steps to follow:

1) Measure and cut the drywall to fit the archway. This will make the drywall more flexible since you'll be cutting it into a narrow strip.

2) Spray water on both sides to dampen the drywall.

3) Prebow the drywall to a fairly tight radius. Once you've done this, you'll be able to bend it the rest of the way to fit the arch. Just bend it by hand and force it into the archway.

4) Fasten the drywall into place. When the drywall dries, it will regain its original stiffness.

Archway before wallboard installation
Figure 3-20A

Archway with wallboard installed
Figure 3-20B

Figure 3-20A shows an arch before the drywall is installed on the inside surface. Figure 3-20B shows the arch after the drywall is installed.

Sometimes the curve of an archway requires a tighter bend than you can get with prebowing. You can make the bend tighter by scoring the backing paper at close intervals. First measure and cut the strip of gypsum panel to fit the archway.

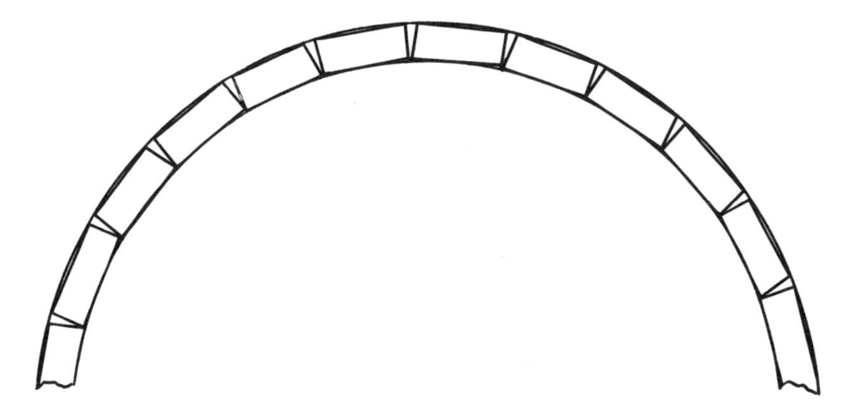

Scoring wallboard to fit tight bend in archway
Figure 3-21

Then use your knife to score the backing paper at intervals of about 1 inch. Do this from one edge of the strip to the other. As you bend the panel, it will break at each of the cuts in the backing paper. If you need to fit a very tight bend, you may need to make the cuts closer than 1 inch apart.

The resulting piece of drywall will have a series of flat sections held together by the facing paper. You'll be able to curve the drywall strip to fit the archway, as shown in Figure 3-21. The surface will appear uneven when you first install this strip. Use plenty of joint compound to smooth out the surface when you finish it.

When you're installing scored, prebowed drywall, you'll need to use more fasteners and will space them closer together. Even though the arch is somewhat self-supporting, the drywall will be rigid only across each flat section.

INSTALLATION GUIDELINES

We've talked about proper handling and storage of drywall materials. And we've learned the proper way to measure, mark and fasten drywall panels. Before we begin our step-by-step sample installation, let's look at six important installation guidelines.

Load the roof before installing any drywall panels: Before you install drywall on either the ceiling or the walls, make sure the roof has been *loaded*. The shingles or other roofing materials should be sitting evenly spaced on the roof, even if they haven't been installed yet. Shingles and other roofing materials are heavy. This additional weight affects the structure of the building, compressing the framing slightly. Gypsum drywall can take some compression without showing the effects. But too much shifting and you'll see some signs of stress at joints. That's why drywall installation has to wait until the roof is loaded.

Install ceiling panels first: This allows you to butt wall panels up against the ceiling panels, giving you the best joint between the walls and ceiling. It's also easier to install the ceiling panels first. Trying to hold a ceiling panel in place while butting it up against a wall panel is possible, but difficult.

The fit of a ceiling panel against the framing doesn't have to be precise. The wall panels will cover any small gaps between the ceiling panels and the framing. Wall panels also give support to the edges of ceiling panels and prevent sagging at the joints.

If you butt the ceiling panels up against the wall panels and the joints do sag, the joint tape will stretch and possibly tear.

Ceiling-to-wall joint and square opening with wallboard trim
Figure 3-22

This results in joints that are uneven and unsightly. And you'll have to repair them.

Figure 3-22 shows a ceiling-to-wall joint and a square opening with drywall installed.

Don't force drywall panels into position: Cut your drywall panels so they fit loosely. You shouldn't have to force them into place. If the fit is so tight that you have to force the panels, they'll probably buckle or bow soon after you're finished.

Avoid butt joints between panel ends: Since panel ends don't have tapered edges, they're hard to conceal. When it's absolutely necessary to have butt joints between panel ends, locate them as far as possible from the center of a wall or ceiling. This will make them less noticeable.

Don't fasten drywall panels to heavy timber framing: Where a drywall panel spans headers or other wide timbers, don't fasten panels to these framing members. The wood will shrink as it ages, pulling the drywall with it and leaving an uneven drywall surface. Instead, attach panels to the edges of framing members that are next to the headers and timbers.

Special procedure for fastening drywall to steel studs: Before you begin fastening drywall panels to steel studs, make sure the studs are all installed in the same direction. Steel studs have a C-shaped cross section, as shown in Figure 3-23. You'll attach the drywall panels to the short section of the "C."

For the drywall joints to come out flat, you must attach panels to the studs in the proper sequence. Fasten the edge of the first panel to the unsupported part of the flange of the stud, as shown in Figure 3-23. Fasten the edge of the second panel to the supported part of the stud flange.

If you install panels in the wrong sequence, the fasteners on the second panel will deflect the flange. This will force panel edges out of alignment.

SAMPLE INSTALLATION

Now let's take a step-by-step look at a sample drywall job. We'll begin by installing the ceiling panels. Next we'll hang the wall panels. Finally, we'll discuss how to strengthen panel joints.

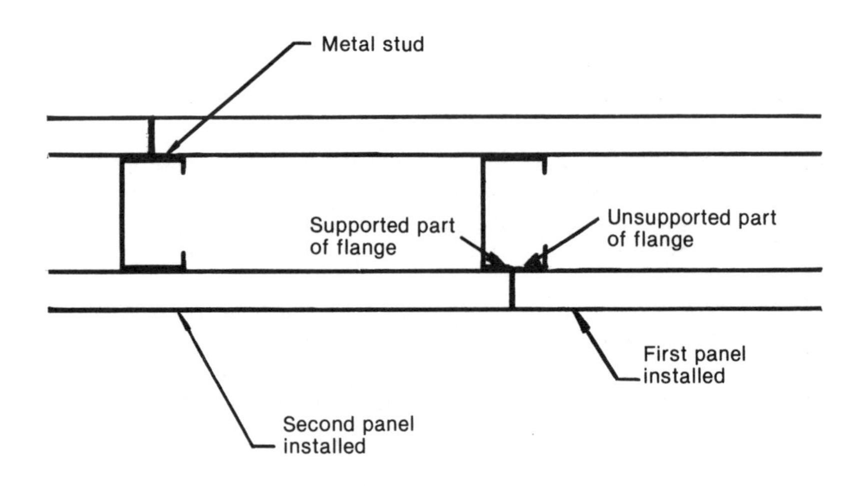

Fastening wallboard panels to metal studs
Figure 3-23

Ceiling Panels

Installing ceiling panels is a two-step procedure. First you hoist the panels into position. Then you fasten them in place.

Hoisting the panels into position: Normally you need two people to hoist a ceiling panel into position and hold it there while you install the fasteners. But if you're working alone, here are three devices that can help you.

T-shaped device— You can hold up one end of the panel with a T-shaped brace. One end rests on the floor and the other holds the panel up against the ceiling. When one end is in place, lift the other end to where you want it. Start by placing one end of the panel on the top of a stepladder. Then lift the other end to a point where you can fit the T-shaped device under it, and wedge the panel against the ceiling. The disadvantage of this method is that it's hard to position the panel precisely. But it may be the only choice if you're working alone.

Lifting device— You can also use a lifting device like the one shown in Figure 2-16 of Chapter 2. This device will hold the panel while you hoist it into position. These lifting devices

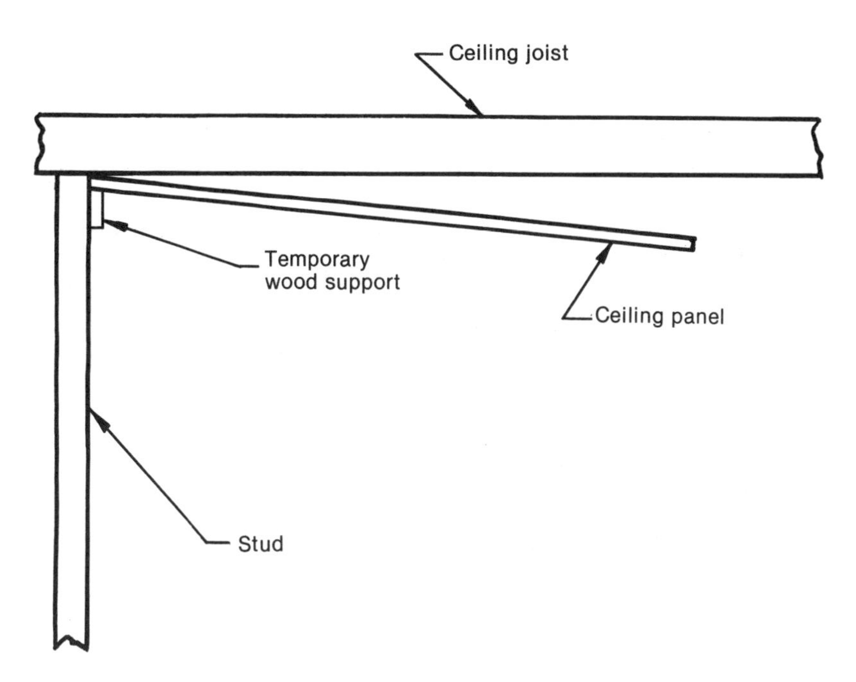

Temporary ceiling support
Figure 3-24

normally have casters. Mount the panel on the lifting device first. Then move it into position. Finally, hoist the panel up against the ceiling.

When you're positioning ceiling panels, be sure the panel ends butt together over a framing member. This is so you can fasten both panel ends to the framing, unless you'll be using back blocking. Back blocking is covered later in the chapter.

Temporary wood support— Wall framing will be exposed when you're hanging ceiling panels because ceiling panels should be hung before wall panels. Where ceiling panels butt against a wall, nail scrap lumber to the studs as a temporary support for those panels. You can then rest one end of the panel on the "shelf" as you hoist the other panel end into position. Here's how to set up the temporary support shown in Figure 3-24.

1) Measure down from the bottom surface of the ceiling framing and place marks on the wall studs at the thickness of the ceiling panels plus about 1/8 inch.

2) Find a piece of scrap lumber that is straight on one edge. It should be at least 1 x 2-inch stock and should be long enough to span at least two studs. Hold the scrap horizontally against the studs and mark the location of the center of each stud on the piece of scrap lumber. Drill 1/8-inch holes at each mark.

3) Using your level and two drywall screws, attach the piece of scrap lumber to the studs. Make sure the straight edge is up. The top surface of the scrap lumber should be even with the marks you made on the studs.

Rest one edge of the ceiling panel on this temporary shelf as you hoist the panel into place and install the fasteners.

Fastening the panels: When you install the fasteners, start at the center of the panel and work out toward the edges. This keeps the panel from bowing out in the center. If the center of the panel bows, you won't be able to hold it tight against the joists. Sagging panels make your finished ceiling look like a tent roof.

Be sure to hold each panel tightly against the ceiling joists when driving fasteners. A panel that isn't flush against the joist has no support directly behind the panel. If there's a gap between the back of the panel and the joist, your hammer can knock a hole in the panel before the nail is fully driven. If you're lucky, the nail or screw will just tear through the facing paper. This weakens the holding power of the fasteners. Facing paper helps spread the load of the fastener head over a larger area. If you tear the facing paper, install another fastener about 1½ inches away from the one that tore the paper.

Make sure your fasteners are no closer than 3/8 inch to any edge of a drywall panel. When fasteners are installed closer than that, the edge may crumble or break at the fastener location. You need enough drywall surrounding the fastener to provide a good bearing surface for the fastener head.

Nails— When using nails to fasten the drywall, the final hammer blow must drive the nail slightly below the surface plane and leave a small dimple in the drywall. Joint compound is spread in this depression to conceal the nail head. The dimple should be no more than 1/32 inch deep. This depth won't tear the facing paper of the drywall. See Figure 3-25.

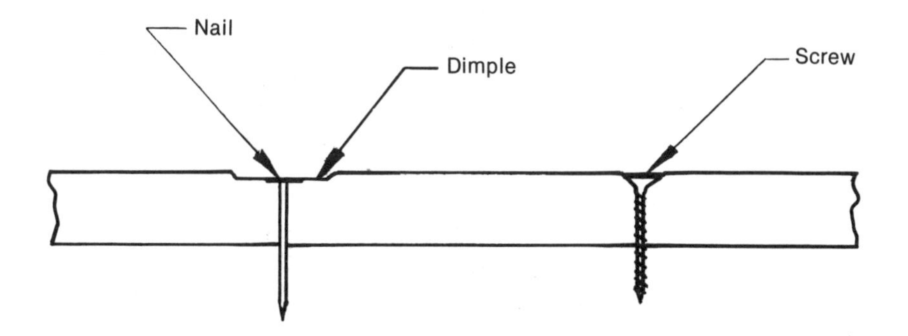

Nail and screw installation
Figure 3-25

The nailing procedure is different for veneer-base drywall that will be coated with a thin layer of plaster. In this case, drive nails so the heads are just flush with the surface of the panel. There's no need to conceal nail heads with joint compound.

Screws— When using screws to fasten the drywall, drive them in far enough so the head is slightly below the surface of the panel, as shown in Figure 3-25. The one exception is, again, when you're installing veneer-base drywall. In that case, the head can remain flush with the panel.

Be sure to adjust the screw gun for the correct depth. Otherwise you'll tear the facing paper. If you're using a variable-speed electric drill, you'll have to be a little more careful about the depth of the screw. Most power screwdrivers turn at about 300 rpm. This is usually slow enough to give you good control over screw depth.

Be careful to drive screws straight. Crooked screws will have one edge of the screw head sticking above the panel face. This makes it impossible to conceal the screw head. Driving a crooked screw far enough to conceal all of the head will tear the facing paper. This reduces the holding power of the screw.

Wall Panels

Once the ceiling panels are in place, install the wall panels. Standard drywall panels come in 4-foot widths. If the ceiling height is more than 97 inches (8 feet, 1 inch), you may want to install panels with the length parallel to the wall studs. This way you can select a panel length that spans from floor to ceiling. Otherwise, it will take more than two panels to span the floor-to-ceiling distance.

When installing drywall panels horizontally (with the length perpendicular to the studs), butt the top panel against the ceiling first. That ensures a good joint between the wall and ceiling. The bottom panel doesn't need to go all the way to the floor. The wall-to-floor joint will probably be covered by some sort of baseboard. When installing panels horizontally, use the longest panels you can. The longer the panel, the less joint finishing.

Where a joint falls near an opening for a door or window, center the joint over the opening. This makes the joint shorter. It also keeps the joint away from the edges of the opening where it may be harder to conceal. Try to keep all joints at least 8 inches away from the edges of openings. This reduces the need for narrow strips of drywall.

When driving fasteners through drywall into the framing, hold the drywall tightly against the framing with your other hand. See Figure 3-26. This ensures a tight fit between drywall and the framing. It also helps prevent surface damage to the drywall.

When you finish installing the drywall, walk around the room pressing firmly against the wall panels and the ceiling panels. Look for any sections of board that aren't tight against the framing. Renail any areas that aren't tight. When you're driving nails or staples, the pounding may loosen some fasteners. Also check adjacent rooms for loose fasteners.

Strengthening Panel Joints

Ridging— is a surface defect at end joints. It's caused when the ends of two panels don't meet at exactly the same plane. Ridging in ceilings is particularly noticeable because interior lighting creates shadows behind any irregularity in the surface. Reduce the possibility of ridging by installing back blocking behind end joints in ceiling panels. An alternative to back blocking is to use strip reinforcement at panel end joints. Let's take a careful look at both back blocking and strip reinforcement.

Back blocking— Back blocking consists of pieces of drywall cut to size and placed between ceiling joists. Laminate back

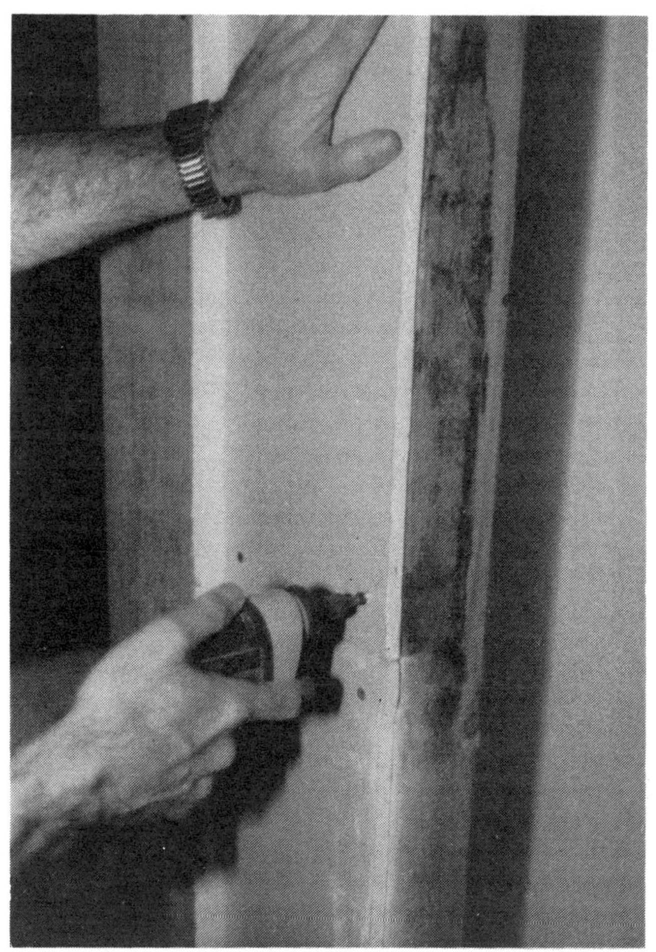

Installing wallboard panels
Figure 3-26

blocking to the back side of the ceiling panels to provide reinforcement at the joints. To do this, you'll have to position the ceiling panels so that end joints occur between ceiling joists. Install the panels perpendicular to the joists.

Figure 3-27 shows a back-blocking installation. To install back blocking, follow these five easy steps:

1) Install the first ceiling panel so the end joint is between joists.

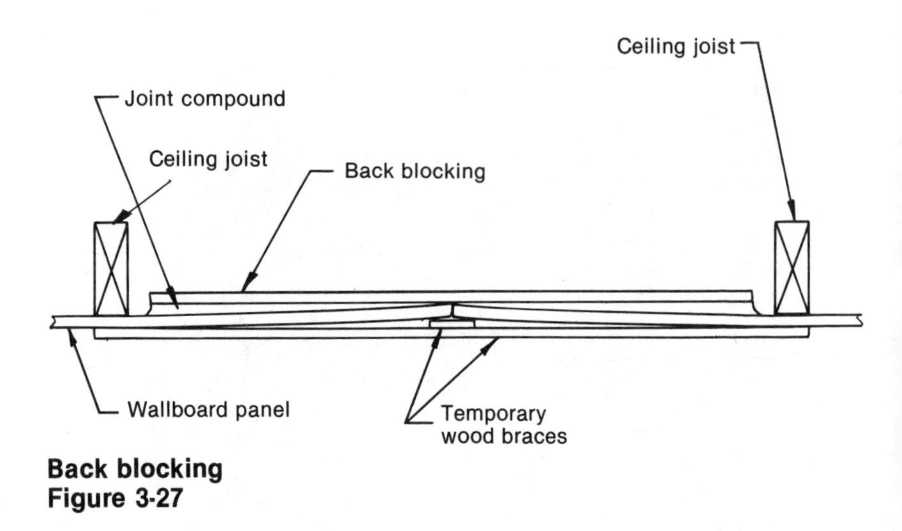

Back blocking
Figure 3-27

2) Cut pieces of drywall to fit between the joists and to extend the full width of the first ceiling panel.

3) Use a notched spreader to apply the joint compound. The spreader should form beads that are 1/2 inch high, 3/8 inch wide and are spaced about 1½ inches on center along the face of the back blocking.

4) Slip the back blocking, face down, over the end of the installed ceiling panel. The joint compound will act as an adhesive between the back blocking and the ceiling panel.

5) Install the next ceiling panel.

You may want to install temporary wood braces along the joint and hold them in place with wood strips placed at right angles. This will help force a tapered joint where the back blocking is installed. Look again at Figure 3-27. You can remove the temporary braces when joint compound on the backing blocks has dried.

Strip reinforcement— If you don't want to install back blocking, use strip reinforcement to strengthen the joints and prevent ridging. Strip reinforcement requires two thicknesses of drywall: base ply and face ply.

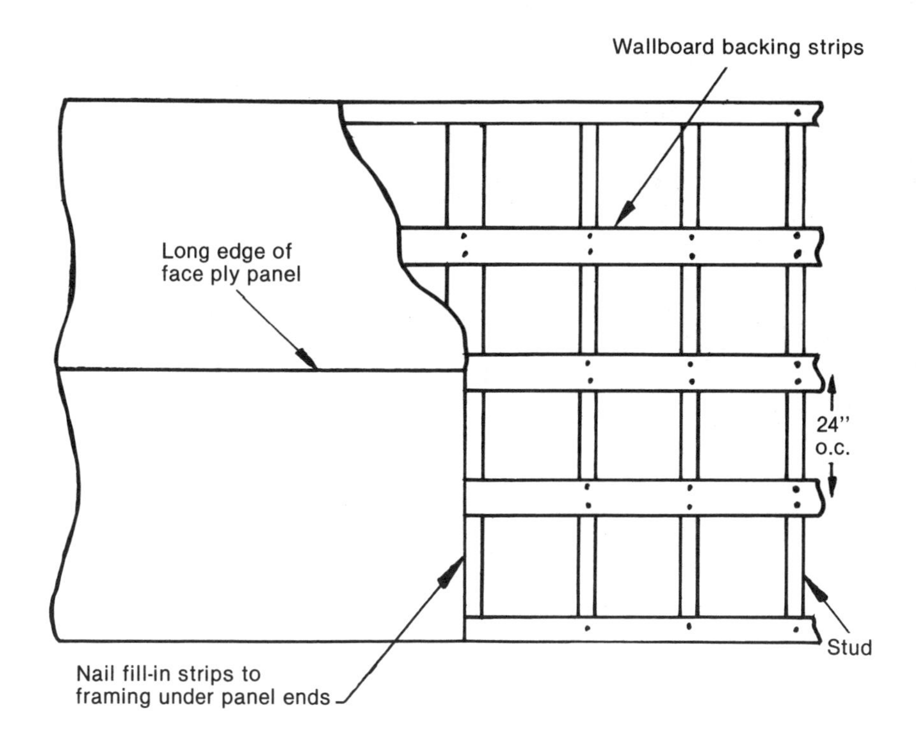

Strip reinforcement
Figure 3-28

The base ply consists of drywall strips, rather than whole gypsum panels. See Figure 3-28. Use scrap pieces of drywall for these backing strips. The strips are normally 8 inches wide and 3/8 inch thick. But they should be no more than 4 inches wide near the floor or ceiling.

Install the strips perpendicular to the framing and space them 24 inches on center. Nail the strips at each framing member. Drive the nails about 1 inch from each edge of the strip.

Position the strips so the long edge of the face ply panel can rest on a backing strip. There will be a gap under the face ply panel end where the framing isn't covered by the strips. You need additional support in this area. To fill in the gap, nail short strips of drywall to the framing members under the face ply panel end joint. Look again at Figure 3-28.

Apply drywall adhesive to the face of both the regular backing strips and the fill-in strips before you install the face ply panels. Since you're using adhesive to laminate the face ply to the strips, you can use fewer fasteners in the face ply. For ceilings, drive nails 24 inches o.c. at each joist. For walls, drive nails 24 inches o.c. at the ends of the face ply panels and along panel edges at each stud.

You'll also need a temporary fastener at every other stud in the center of each panel until the adhesive sets. You can use double-headed form nails or screws as temporary fasteners. They're easy to take out when you're done with them.

If the framing members are spaced more than 16 inches o.c., use 1½-inch drywall screws in the center of the face ply panels to secure the panels to the strips. Install these screws *between* the framing members. This provides additional stiffening. You can use nails for the rest of the fasteners.

Strip reinforcement can give your drywall installation nearly the stiffness of a double-ply installation, but at a lower cost.

MULTI-PLY CONSTRUCTION

There are times when the job will require two or more layers of drywall on ceilings or walls. There are five advantages to multi-ply construction:

- Increased fire resistance
- Greater noise protection
- Base for attaching decorative wall panels
- Greater strength
- Smoother finished surface

The first layer of drywall in multi-ply construction is known as the *base ply*. The layer or layers of drywall installed on top of the base ply are known as the *face ply*.

Base Ply

Here are some key points to remember when installing the base ply in multi-ply construction.

Ceiling panels— Apply base ply ceiling panels with the long edges perpendicular to the framing members. It's O.K. if the ends of the *base ply* panels don't fall directly on the ceiling joists. But the *face ply* panel end joints must. This allows you to fasten the ends of the face ply panels through the base ply panels into the joists.

You should offset the base ply panel joints from face ply panel joints. This adds strength because every end joint will have a solid gypsum panel overlapping it.

Where the framing members are metal, use drywall screws to fasten the base ply to the framing. Where framing members are wood, use nails, screws or staples to fasten the base ply. The technique for installing nails or screws in base ply is the same as the technique for installing them in single-ply construction.

When installing staples, remember to hold the drywall firmly against the framing while you drive the staples. Drive them in far enough so the crown of the staple presses tightly against the drywall. But don't drive it in so far that you tear the facing paper. Fasteners that tear the paper have far less holding power.

The minimum *grip length* for staples is 5/8 inch. The staple should penetrate the framing by 5/8 inch. Refer back to Figure 3-5 to select the correct staple length for each thickness of drywall. Figure 3-29 shows the maximum spacing for fasteners in the base ply of multi-ply drywall construction. Notice that the face ply fastening method affects the spacing of base ply fasteners.

Wall panels— Install base ply wall panels with the long edges parallel to the wall studs, *except* where you're installing decorated panels on top of the base ply, or where wall height exceeds 97 inches. In this case, install the base ply or wall panels perpendicular to the studs.

At inside corners, you'll be installing fasteners only in the base ply. And because there are fasteners in only one layer of drywall, this technique is called a *floating corner* or *floating angle.* Floating corners create joints that resist stress better than joints that have the face ply nailed as well.

When installing wall panels, be sure to stagger the joints of the base ply and face ply. That adds strength and stability. Fastening techniques for base ply wall panels are the same as for base ply ceiling panels.

Base ply fastener	Face ply fastening method	Application	Maximum spacing for base ply fasteners
Nails	Laminated face ply	Ceilings	7" o.c.
		Walls	8" o.c.
	Nailed face ply	Ceilings	16" o.c.
		Walls	16" o.c.
Screws	Laminated face ply (framing 16" o.c.)	Ceilings	16" o.c.
		Walls	16" o.c.
	Laminated face ply (framing 24" o.c.)	Ceilings	12" o.c.
		Walls	12" o.c.
	Screwed face ply	Ceilings	24" o.c.
		Walls	24" o.c.
Staples	Laminated face ply	Ceilings	7" o.c.
		Walls	7" o.c.
	Nailed or screwed face ply	Ceilings	16" o.c.
		Walls	16" o.c.

Maximum fastener spacing for base ply in multi-ply construction
Figure 3-29

Face Ply

The face ply is normally the second layer of drywall. But you may be installing two or more layers of drywall on top of the base ply. In this case, your face ply may consist of several layers of drywall laminated together.

Look again at Figure 3-29. Notice that the maximum fastener spacing for the base ply depends on whether you're attaching the face ply with adhesives or mechanical fasteners. The fasteners must be closer together when you use adhesive to fasten the face ply. That's because there won't be any face ply fasteners to penetrate the framing and add holding power.

To install face ply with adhesive, use a notched metal spreader with 1/4-inch square notches that are spaced 2 inches or less on center. Some spreaders have notches that are 3/8 inch wide by 1/2 inch high. Before you select a spreader, read the instructions on the adhesive container to see if the manufacturer has any special recommendations.

Using the spreader, apply strips of *laminating adhesive* along the edges and in the center of the face ply backing paper or to the face of the backing ply. This is known as *strip lamination*.

Some job specifications require you to apply adhesive to the entire panel surface. This is known as *sheet lamination*. You'll also use a notched applicator for spreading adhesive in sheet lamination.

Other jobs require you to use liquid *contact adhesive*. Spread this type of adhesive with a roller that has a short nap. Let each surface dry to the touch before you join them. Contact adhesive is just that. Once the two surfaces come in contact, they are glued. So be careful to position the panels precisely before bringing them together. Press the face ply panel tightly against the base ply panel. Then install the fasteners to hold it in place. These fasteners are installed in the same manner as for the base ply.

Self-Supporting Partitions

You can make self-supporting partitions by laminating drywall panels to a gypsum coreboard. The coreboard can be continuous or in separate sections 6 to 8 inches wide. Spacing of coreboard sections will be given in the job specifications.

Self-supporting partitions use no wood or metal framing, except for metal runners at the top and bottom and at the intersection with another wall or partition. These runners stabilize the partitions. The runners may be channels or two pieces of metal angle.

Figure 3-30 shows a self-supporting partition with tongue and groove gypsum coreboard. Figure 3-31 shows a self-supporting partition using coreboard spacers at intervals between the gypsum facing panels. In both types of construction, you normally install the panels vertically. Depending on the job requirements, you may build self-supporting partitions with single-ply or multi-ply facing.

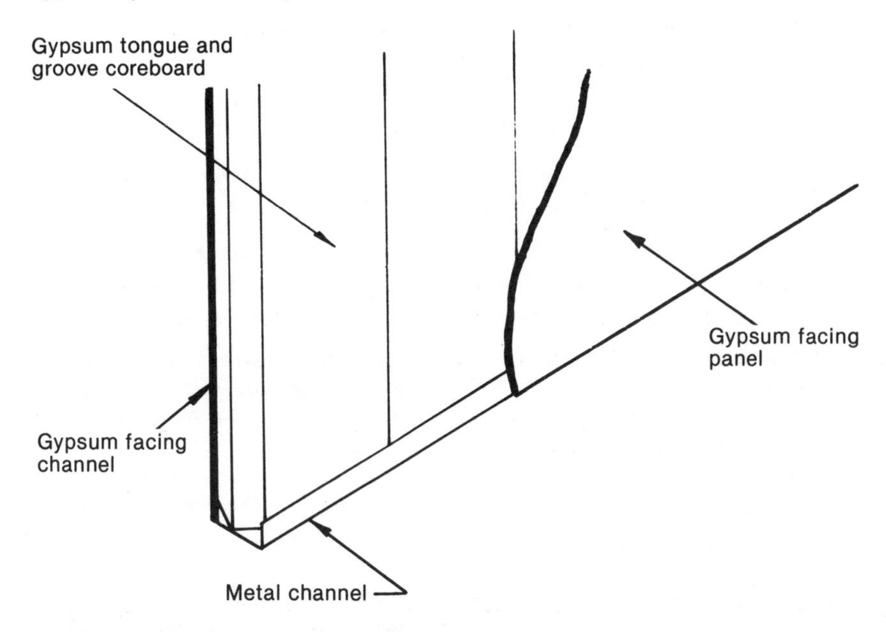

Self-supporting partition with tongue and groove coreboard
Figure 3-30

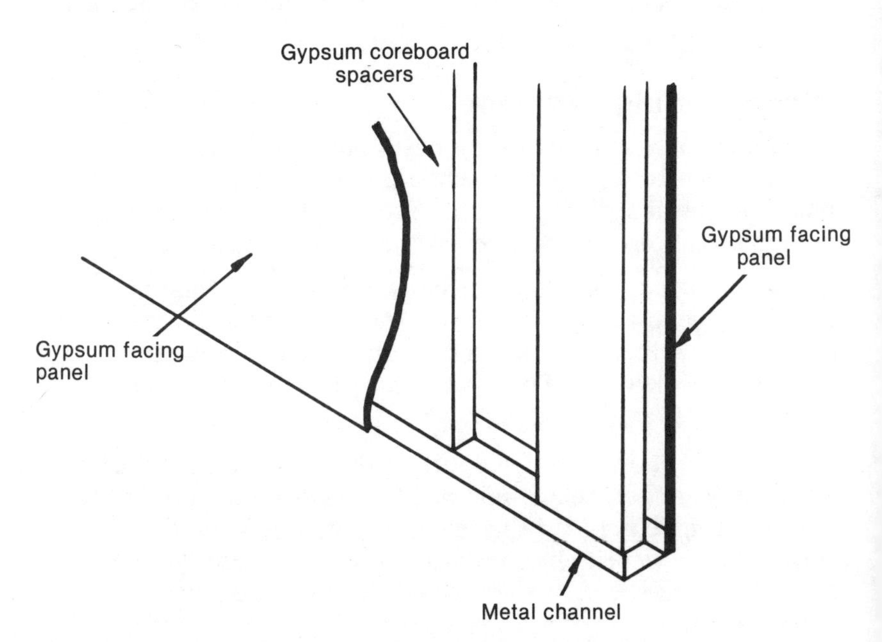

Self-supporting partition with coreboard spacers
Figure 3-31

Here are nine important points to keep in mind when installing self-supporting partitions:

1) Runners can support the gypsum core, the core and one facing panel or both facing panels. But it's essential that you position the runners accurately. Use a plumb bob to position the runners at the top and bottom so the partition will be plumb.

2) Runners that must be installed between ceiling joists should be attached to blocking placed between the joists. Runner is attached to the blocking.

3) The fasteners you use to attach the runners depends on what the runners are attached to. Use screws or nails to attach runners to wood framing. On concrete, use concrete anchors or concrete nails. Use toggle bolts or molly bolts to attach runners to a suspended ceiling.

Install the fasteners 2 inches from each end of a runner. Intermediate fasteners should be no more than 24 inches apart. If you're using channel for top and bottom runners, make sure there's enough room to slide the gypsum panels into the channel.

4) You may want to use channel only at the top or bottom and use angles on the opposite end. This allows you to set one end of the panel in the channel and swing the other end into place against the angle. Then install the second angle.

5) At corners, butt the runners together instead of mitering them.

6) When the coreboards are in place, secure them to the runners using drywall screws. Drive the screws 12 inches on center.

7) Be sure to offset the face ply joints from the joints in the coreboard by at least 3 inches. If you're using spaced sections of coreboard, the sections should span the joints in the face ply. The gypsum coreboard comes as solid sheets with V-shaped tongue and groove edges that fit together, or as sheets that are scored to snap off at 6- or 8-inch widths. These widths are installed with spaces between them. They become, in effect, gypsum studs. They're installed no more than 24 inches on center.

8) Apply beads of adhesive to the coreboard to laminate the facing plies to the gypsum core. Use drywall screws to secure the face plies to the coreboard while the adhesive sets. Space these screws a maximum of 36 inches on center.

9) For greater fire resistance and sound insulation, you can separate the structure (coreboard) behind the face plies in either solid or spaced coreboard sections. If you're using spaced sections of coreboard, double the number of sections and laminate every other section to one face panel and the remaining sections to the opposite face panel. This effectively isolates one face panel from the other. If you're using solid coreboard, set two runners parallel to each other and separate them by the gap specified in the job specifications.

SPECIAL DRYWALL CONSTRUCTION

Sometimes drywall is installed to meet special building requirements. For example, drywall can be used to reduce sound transmission, to meet fire code requirements, add moisture resistance or improve the insulation value of a wall or ceiling.

But using the right materials usually won't be enough. You have to hang the board correctly, apply the right joint treatment, then apply the prescribed finish. Otherwise, the finished job won't have the properties the designer intended.

This section describes the most common requirements. But standards and materials change. And some manufacturers make products that require special treatment. When using materials designed for special applications, always follow the manufacturer's recommendations.

Sound Insulation

The American Society for Testing and Materials (ASTM) has developed standards for testing sound transmission through structures. ASTM Standard E 492 describes how to measure noise impact in floors. ASTM Standard E 90 tells how to measure sound transmission through building partitions. ASTM Standard E 413 uses a standard sound transmission curve to help determine the class of sound transmission for your particular construction project. ASTM Recommended Practice E 497 provides guidelines for controlling sound.

As sound travels through air, it causes anything it meets to vibrate. This vibration transmits sound. To reduce sound transmission, designers eliminate the *direct pathway* sound is using. When the drywall panels in one room connect directly to the drywall panels in the next room, this is a direct pathway. To reduce interior noise levels, designers do two things. First, they try to remove all direct paths for sound transmission through the structure. Second, they make the building as airtight as possible.

Here's an example. Most homes have drywall fastened to both sides of every stud. That type of construction provides a direct pathway for sound. Sound moves directly from the air in one room to the drywall to the stud under the drywall to the drywall on the other side of the wall and finally to the air in that room. The result is that voices in one room can be heard clearly in the next room.

Sound transmission like that isn't a serious problem in most homes. Even if the walls provided better sound insulation, noise would still travel down the hall and through open doorways. Anyhow, the Jones family is used to their own noisy kids. But it's a more serious problem when noise is traveling from one apartment or condominium to another. The Smiths don't appreciate the Jones' noisy kids and don't want to be quite so well informed about the Jones' family affairs. Like any neighbor, they wish the walls offered better sound protection. Fortunately, many building codes and more progressive builders are providing that protection.

Here are six ways to reduce sound transmission in your drywall installations:

Offset studs— You can stagger studs so drywall on each side of the wall is fastened to a different set of studs. Do this by installing 2 x 4 studs on a 2 x 6 plate, as shown in Figure 3-32. By staggering the studs, there's no direct contact between the two drywall panels. This eliminates the direct path for sound transmission.

Parallel walls— You can build two 2 x 4 walls that are parallel to each other but that don't touch each other. See Figure 3-33. Again, since the two walls aren't touching each other, there's no direct path for sound transmission.

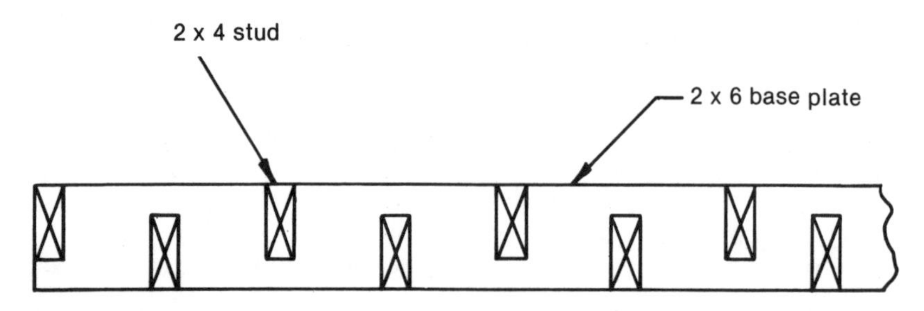

Offset studs
Figure 3-32

Insulation— Adding insulation between drywall panels increases sound insulation even more. Fill the spaces between the framing members with sound-absorbing material. Fiberglass insulation batts will decrease sound transmission.

To be effective, the batts should fit snug in every cavity. You'll have to cut the batts carefully so they fit around electrical boxes, pipes and wiring. If you're using insulation batts that have a vapor barrier (foil or kraft paper on one side), staple paper flanges of the vapor barrier to the framing. If you use unbacked fiberglass batts, cut them so they're held in place by a friction fit.

If vapor barrier is included in the wall, it should be on the inside, or occupied space side, of the insulation. For instance, in a wall between a house and the garage, vapor barrier should be on the house side of the wall.

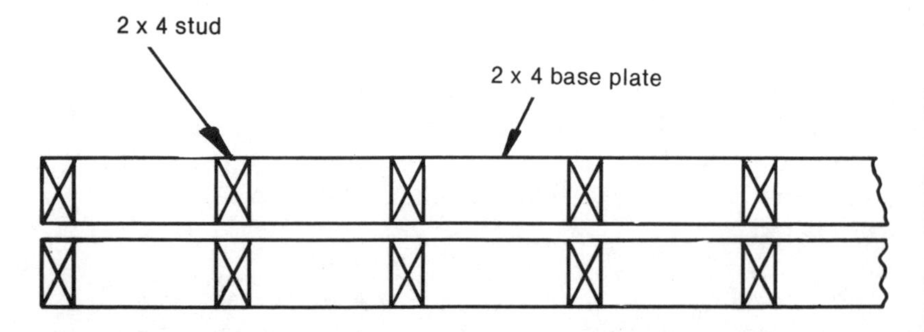

Parallel walls
Figure 3-33

There's an important reason for this. Warm air in the house holds more moisture. This moisture will condense and form droplets of water when exposed to a cold surface — like water condensing on a glass of iced tea on a hot August afternoon. If moisture condenses on the vapor barrier, you want that moisture on the house side of the wall, not in the insulation.

The vapor barrier should keep the fiberglass batting from getting wet from condensation. This is important for two reasons. First, wet fiberglass batting is a very poor insulator. Second, the fiberglass doesn't dry out very quickly. Instead, it holds moisture in the wall, making decay more likely.

Resilient channels— You can also use sound-deadening board and resilient clips or resilient channels to eliminate the sound transmission path.

Sound-deadening board is available as a wood-base fiberboard, plastic foam board or fiberglass-based board. It comes in thicknesses from 1/4 inch to 5/8 inch. Use the wood-base or plastic foam board only where combustible materials are acceptable. If noncombustible construction is required, use the fiberglass-base sound-deadening board over metal framing.

Install the sound-deadening board parallel to the framing. Then install resilient channels perpendicular to the framing members, as shown in Figure 3-34. Resilient channel has a flange along one edge. The flange goes down. When you install the flanged edge of the channel on the bottom, it acts as a sort of hinge. When the drywall is attached to the channel, its weight will pull down on the channel. As the channel is fastened only along its bottom edge, the top edge of the channel (which is not fastened to the studs) will move slightly away from the studs. This will minimize the contact between the drywall and the studs, further diminishing sound transmission through the wall.

Install the resilient channels with screws spaced 24 inches on center, except when you're installing 1/2-inch thick drywall with a single-coat veneer plaster finish. In that case, space the resilient channels 16 inches o.c. Install the first channel 2 inches up from the floor. The top channel should be about 6 inches down from the ceiling.

If you use nails, they may work loose from the framing as the framing lumber dries out, causing the drywall to sag. This is especially true of ceiling installations. If the ceiling is suspended, rather than framed with ceiling joists, use wire to attach the channels to suspended steel members. You can use resilient clips

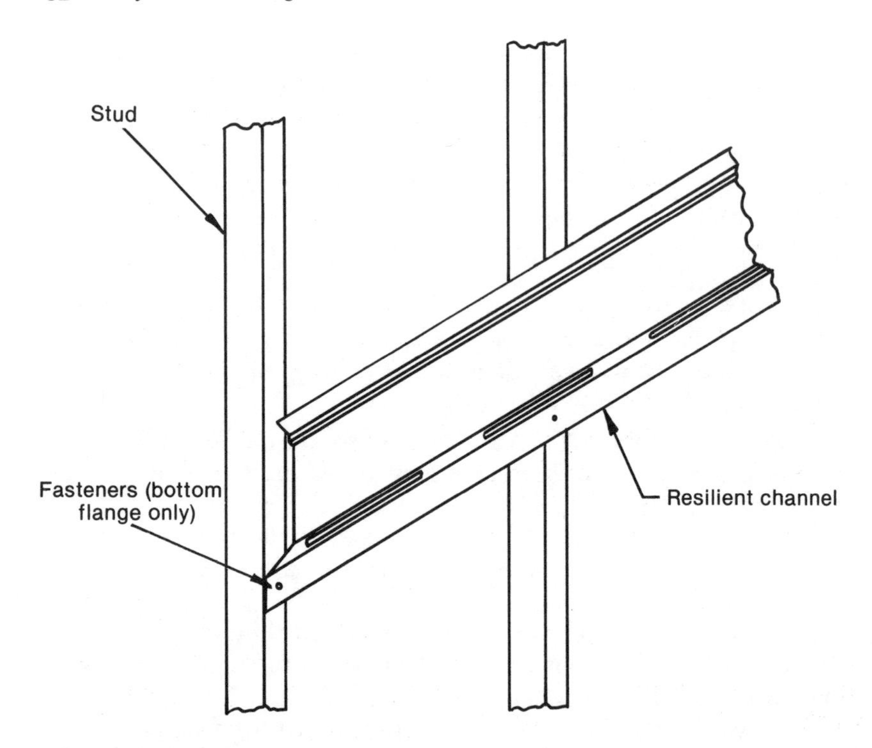

Resilient channels
Figure 3-34

to attach the base ply in multi-ply wall installations. But clips aren't strong enough for ceiling installations.

Where you have to splice the channel, do it over a framing member. Don't butt the channels together. Place the first channel across the framing member. Lay the channel to be spliced over the first channel so they nest together. Drive a fastener through both channels into the framing member.

You can use resilient channels for walls and ceilings, but they can support only limited weight without losing their effectiveness. That's why you can't use them to support more than a single ply of drywall. When using resilient channels on ceilings, install them perpendicular to the ceiling joists. Where double-layer drywall construction is required, first attach the base ply to the framing. Next, install the resilient channel on the base ply. Finally, fasten the face ply to the resilient channel.

The disadvantage of resilient channels is that you can't hang fixtures or cabinets from wall panels mounted this way. The added weight of the fixtures would press the drywall to the framing, compressing the resilient mounting. Even the fasteners you'd install to support fixtures or cabinets would compress the resilient mounting too much. That leaves a direct path for sound transmission.

Caulking— Once you've installed the drywall, caulk the entire perimeter of the wall partition to make it as airtight as possible. If there are any open airways through the wall or around the drywall panels, sound will leak in.

Make the cutouts for electrical boxes as small as possible. Leave a minimum gap around the box. Then caulk that gap with a resilient (nonhardening) caulking compound to provide an airtight seal. This eliminates a direct sound path.

Fixture placement— Avoid back-to-back placement of fixtures that are recessed into the wall. Fixtures can transmit sound through the wall. And avoid routing heating or air conditioning ducts through walls that are intended to isolate sound. Duct work is a good sound transmitter.

Using Fire-Resistant Drywall

ASTM Standard E 119 tells how fire resistance ratings are established. Walls, ceilings, columns and beams are tested under ASTM Standard E 119. Here's how the test works.

The temperature of the wall surface being tested is raised at a standard rate to a specific temperature, simulating what would happen in a fire. The temperature on the *unexposed* side is then measured. To pass the test, the wall must be of a certain minimum size and must be able to hold up certain loads when heated. For instance, if it is a load-bearing partition, it must be tested with weight applied to simulate the load it would normally carry.

Drywall is noncombustible, so it adds to the fire resistance of any structure. It insulates wood studs from the heat of a fire for a period of time. Type X drywall has fiberglass reinforcement to increase its fire resistance.

Regardless of the type and thickness of drywall installed, the wall is still considered "combustible" as long as wood studs are used in the wall. The wall is termed "noncombustible" only if *none* of the materials in the wall will burn. In combustible construction, the four common fire resistance ratings include: 1-hour, 2-hour, 3-hour and flame-spread ratings.

1-hour rating— When you install 5/8-inch thick Type X drywall on both sides of wood studs, the wall will have a 1-hour fire rating *if the drywall is properly installed.* Proper installation includes taping all of the joints and applying joint compound, so there's no path for the fire to follow around the drywall.

2-hour rating— If you need a 2-hour fire rating, install two thicknesses of 5/8-inch thick Type X drywall on each side of the wall. Be sure to stagger the drywall joints. Also stagger the joints on opposite sides of the wall so they don't match. This minimizes the possibility of fire finding some path through at the panel edges.

3-hour rating— For a 3-hour fire rating, install three layers of 5/8-inch thick Type X drywall on each side of the wall and fill the spaces between the studs with fiberglass insulation. This reduces heat transfer to the studs and through the wall.

Flame-spread rating— This rating applies to finish materials. It measures how fast and how far flames will spread over surfaces under specific test conditions. ASTM Standard E 84 tells how to determine flame-spread ratings.

Your local building code probably specifies flame-spread ratings for interior finish materials in public gathering areas, exits and corridors. Be sure to check these code requirements. Flame-spread ratings range from 0 to 100. Gypsum plaster has a flame-spread rating of 0. Oak has a flame-spread rating of 100. The rating for drywall is 10 to 15.

Fireproofing columns— Columns are fireproofed so the building doesn't collapse in a fire. Even steel columns can bend and sag from the heat of a fire. Drywall makes good fireproofing for building columns and offers another advantage. When installed, the drywall is ready to finish. You can apply two or more thicknesses of drywall to columns. Here's how to do it.

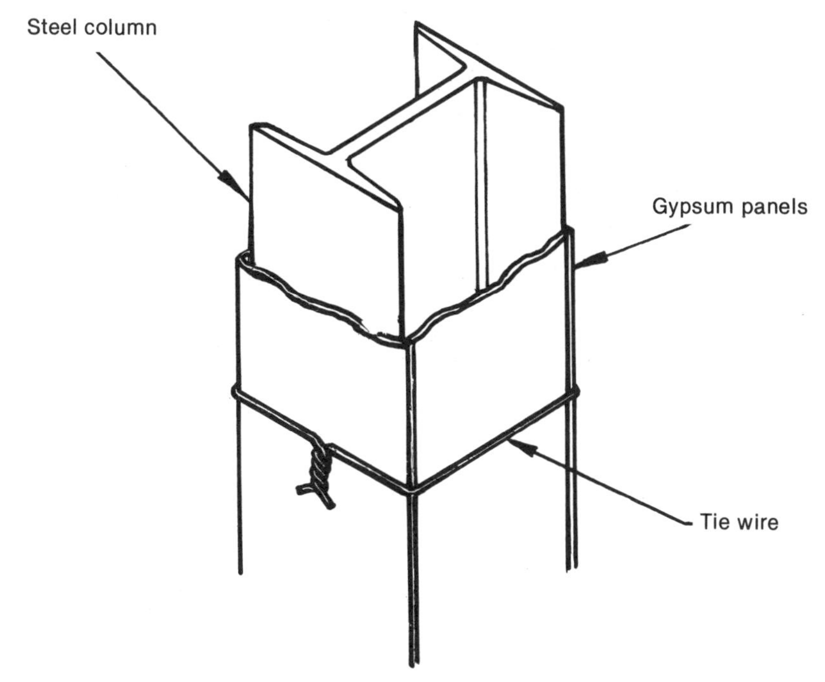

Fireproofing columns
Figure 3-35

Cut pieces for the base layer to fit around the column on all sides. Secure it with strands of steel wire wrapped around the column. Twist the ends tightly, as shown in Figure 3-35. An alternative way is to secure the base layer to steel channels that are spot-welded to the column. Use screws to fasten the drywall to the steel channels. Then use adhesive and screws to fasten the next layer(s) to the base layer.

Another method for fireproofing columns is to apply metal lath to the columns and gypsum plaster on top of the lath.

Moisture-Resistant Installations

Standard drywall is not moisture-resistant. It will crumble if exposed to moisture long enough to soak the drywall. You can use standard drywall in applications where there is *limited* exposure to moisture, provided the drywall is properly protected by tile or oil-base paint.

In Chapter 2, I mentioned some specially formulated products that have much better moisture resistance than standard drywall. If you're installing drywall in a *high-moisture* area, use panels that are specially designed to resist moisture. Laundry rooms and bathrooms are typical high-moisture areas.

Interior applications— In bathtub and shower areas, the shower pan or bathtub must be in place before you install the drywall around it. Install temporary 1/4 inch spacers around the top of the shower pan or tub to locate the drywall panels. You'll fill this space with sealant after the tile is installed.

Use at least 1/2-inch thick drywall on studs spaced no more than 16 inches on center. At spacing more than that the drywall would be too flexible, particularly where it joins a tub or shower pan. Where studs are spaced more than 16 inches o.c., install blocking between the studs. Space the blocking 4 feet o.c. Install the blocking just above the top rim of the tub or shower and then up 4 feet. Make sure the blocking supports the edges of horizontally installed drywall between the studs.

As an alternative to blocking, you could use 5/8-inch thick water-resistant drywall. Be sure to seal the cut panel edges with a coating of water-resistant joint compound.

You can use adhesive to fasten the drywall as long as the tile you're installing is 5/16-inch thick or less. Thicker tile is too heavy for drywall fastened with adhesive. If the tile is thicker than 5/16-inch, install the drywall panels with nails spaced 4 inches o.c. or screws spaced 8 inches o.c.

Once the drywall is fastened in place, tape the joints and fill them with water-resistant joint compound. Seal all penetrations through the drywall with water-resistant joint compound or caulk. I like to use adhesive-backed fiberglass mesh joint tape. It's water-resistant and easy to apply.

Work carefully when sealing and finishing the joints. The joints should be as moisture-resistant as the panels themselves. Seal the drywall surface with a water-resistant sealant before applying the tile. You won't need a third coat of joint compound if you're applying tile to the drywall surface.

Exterior applications— Sometimes drywall is used on exterior surfaces. For example, drywall can be used under eaves and for the ceiling of a carport. Standard drywall is acceptable in these locations, as long as the panels aren't directly exposed to the weather and they're properly finished with an exterior alkyd or oil-base paint. But it's better to use moisture-resistant drywall panels.

Install fascia boards to at least 1/4 inch below the surface of the ceiling panels. The fascia will form a drip edge to keep water off the drywall panels. Any time you install drywall outside, make sure that there will be no water runoff across the panels.

Protect exterior drywall edges with metal edge trim. And keep the drywall edges at least 1/4 inch from all masonry surfaces. Drywall edges left unprotected or in contact with masonry will soak up water like a sponge. No drywall will last long under those conditions.

In ceiling installations, use expansion joints in the following locations: where there are expansion joints in the roof; where the legs of "U," "L" or "T" shaped ceilings come together; and in runs that are more than 30 feet long. Leave a space 1/16 inch to 1/8 inch between panel ends. All expansion joints should fall over a stud or joist.

Exterior gypsum sheathing, as described in Chapter 2, is used for the exterior walls of buildings. It's never used as the final building siding, but forms a sturdy, fire-resistant backing for many different types of siding, including aluminum and masonry.

Install exterior tongue and groove gypsum sheathing horizontally with the tongue edge up. Where there's no diagonal bracing in the framing, rest the bottom edge of the drywall panel on the subfloor. Space the perimeter nails 4 inches o.c. Space *field* nails 8 inches o.c. (Field nails are all nails except those around the edges of panels.) Apply adhesive along external corners. If there's diagonal bracing in the framing, you can space all nails 8 inches o.c.

Radiant Heat Installations

Electric heating panels provide *radiant heat* for a room. As electric current passes through the wire elements, the panels get warm. The entire surface of the wall or ceiling radiates heat into the room. Electric heat panels are usually used on ceilings rather than walls.

You can buy factory-made electric heating panels with wires already embedded in the drywall core. These usually arrive from the factory ready for installation and connection to a power source. Follow the manufacturer's instructions when cutting, hanging, taping and finishing these panels.

You can also install the same type radiant heating wires between layers of standard drywall. Here's how. First, use nails, screws or staples to install gypsum backing board on the wall or ceiling. Refer back to Figure 3-14 for fastener spacing requirements. Then attach the heating cable to the backing board, following the cable manufacturer's recommendations and job specifications. Make sure you draw the cable up taut so it doesn't sag. Sagging cable is hard to cover.

Install the cable so it crosses over the joists or studs at a point 4 to 6 inches from the edge of the wall or ceiling. Leave a free end that's long enough for connection to the electric service.

When all cable has been installed, it has to be tested and inspected. *The cable will be inaccessible once the installation is complete.* Maximum cable temperature is 125 degrees F. When the temperature increases beyond 125 degrees F, it becomes a fire hazard and can also suck the moisture out of the drywall or filler plaster.

Next, filler plaster is applied over the cable. Special filler plasters are made that promote heat transmission from the cable to the room. Apply plaster so it completely covers the cable. Smooth the plaster to a flat surface. You may want to install wood furring parallel to the cable runs. This will give you a screed to use when leveling the plaster. Kiln-dried furring is best.

Level the filler plaster so the face ply of drywall will make complete contact with it. Any gaps between the plaster and the face ply will act as insulation, decreasing the efficiency of the heating cable.

The filler plaster acts as an adhesive before it sets up. So you can install the face ply with fasteners spaced 16 inches o.c. If you allow the plaster to set up before installing the face ply, add a thin coat of plaster to act as an adhesive and gap filler. In this case, refer back to Figure 3-14 for the correct spacing of face ply fasteners.

After the filler material is dry, you can fill and finish the face ply joints. Be sure the filler material is completely dry before you do this. This allows the face ply to take its final position. Drying time may be from one to two weeks, depending on the temperature and humidity of the surrounding air. Don't turn the heating system on until the joints are completely dry. Otherwise the joints may shrink and crack.

This chapter has covered a lot of ground — because there are so many different gypsum drywall products and ways to apply those products. So far we've looked at how to place the drywall in position with the proper number and types of fasteners. In the next chapter we'll learn how to take the next step in the installation process: finishing the joints.

Chapter 4

JOINT TREATMENTS

The big advantage of drywall is that it goes up fast. Every panel covers a lot of wall area. But a single drywall panel is seldom large enough to cover an entire wall or ceiling. And wherever you install more than one panel, you'll have joints between the panels. Joints have to be covered so they don't show in your finished wall surface.

Most buildings move and shift and contract and expand at least a little. Movement has several causes. If a house is built on adobe soil, for example, the soil will expand when it gets wet and shrink as it dries out. This expansion and contraction can cause the building to move up and down, and the movement probably won't be uniform. This uneven movement puts uneven loading on the foundation that's transmitted to the entire structure, including the drywall.

Another source of movement is changes in moisture content of the framing lumber. As wood dries, it shrinks. This shrinkage is uneven, causing walls and ceilings to move, putting stress on the drywall. The paper covering on drywall panels makes them stiff and strong, adding stiffness and strength to the framing. But the joints are the weak link in the chain, *if* they're not properly done. If your joints are faulty, stress cracks will concentrate at the joints.

Good joints resist cracking and look better longer. But that's just one good reason for making good joints. Fire resistance ratings depend on the strength of both the panels and the joints. A joint that isn't properly sealed will let fire through. Good joints are also essential for moisture resistance and for sound insulation. A good joint can be as strong and durable as the panel itself.

In this chapter, I'll explain the right way to make drywall joints. We'll begin with a look at the materials and tools you'll need. Then we'll examine the joint treatments you need to use.

MATERIALS

A durable drywall joint is a strong joint. If you just filled joints with joint compound, they'd have very little strength. Here's where joint tape enters the picture. When you embed joint tape in the compound, you're extending the strength of the facing paper to the next panel. Joint compound not only fills the joint, it acts as an adhesive to bond the joint tape to the panels. Tape bridges the gap between panels to provide continuity between the facing paper on adjoining panels. Figure 4-1 shows a drywall joint the way it should be.

To get started, let's look at the most common types of joint compound and joint tape.

Joint Compounds

The three types of joint compound you'll use most often in your drywall work are standard joint compound, topping compound and all-purpose compound.

Standard joint compound— You have the choice of either dry or ready-mix formulas. Ready-mix is ready to apply straight from the bucket. Dry compound mix costs less than ready-mix compound. It takes up less storage space and isn't as sensitive to temperature changes as ready-mix compound. But there are also several disadvantages to using the dry mix.

You must have a source of clean water to mix the compound. You'll need a mixing container, mixing tools and the time to mix it. And if you don't get the exact proportions of water and dry mix, the compound won't work properly.

If you decide to use the dry mix, here are four important points to remember:

- If you've been storing the mix at very cold temperatures, bring the mix to room temperature the day before you're ready to use it. That makes the powder easier to mix.

- Follow the manufacturer's mixing directions precisely. It's essential that you get the right proportion of dry mix to water.

- Use clean water. And make sure your mixing tools and containers are clean.

- Mix the powder and water thoroughly. There shouldn't be any lumps in it. There's a compound mixing tool available that looks like a heavy-duty potato masher. In addition to the mixing tool, use a paint-stirring paddle or other stiff stick.

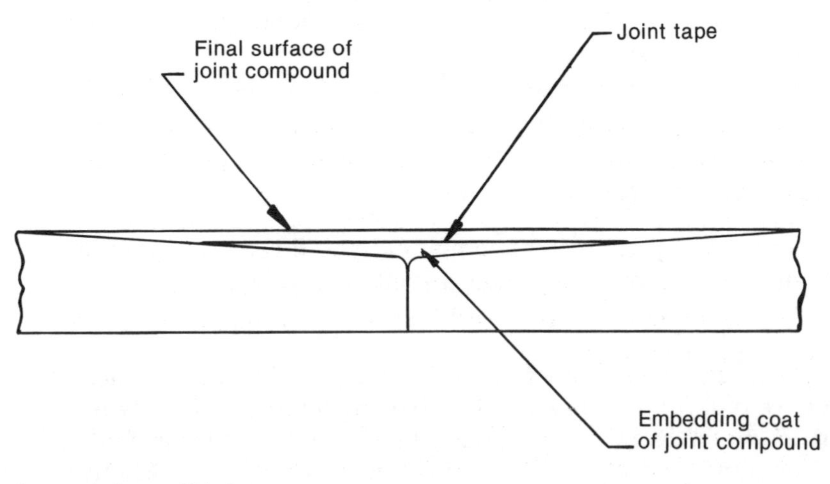

A good drywall joint
Figure 4-1

Topping compound— Topping compound makes a good finish coat for a joint. Because it has a smoother consistency than joint compound, it gives the joint a smoother finish. It's also easier to sand and doesn't shrink as much as joint compound. But don't be tempted to use topping compound for the tape-embedding coat or for the first coat over metal corner bead. Topping compound doesn't have the bonding power of standard joint compound.

When applying topping compound, be careful not to overwork it. Spread it on evenly. If necessary, go over it one more time to feather it properly. But if you smooth it too many times, it'll start to ball up, making the finish rough rather than smooth.

All-purpose compound— This type of compound is a compromise between standard joint compound and topping compound. It doesn't have quite the bonding strength of standard joint compound, but it's suitable for embedding tape. It's not quite as smooth or shrink-resistant as topping compound but it'll do the job. All-purpose compound is particularly useful when you're working on a number of smaller jobs and don't want to stock several types of materials.

Joint Tapes

There are two basic types of joint tape. The oldest and most commonly used tape is made of paper. The second is made of fiberglass.

Paper tape— Paper tape is about 2 inches wide and comes with a crease down the center. The crease is a guide for folding the tape when you use it in corner joints. Paper tape normally has holes in it. The holes range in size from very small holes (that you can see only by holding the tape up to a light) to holes about 1/16 inch in diameter. The holes increase gripping strength when joint compound is forced through them in the embedding process.

Fiberglass tape— This type of tape has fiberglass strands woven into a coarse mesh. It's about 2 inches wide, is easy to apply, and is moisture-resistant. It's good for high-moisture areas where you're using moisture-resistant drywall.

Fiberglass tape is available with adhesive on one side, but it costs about twice as much as paper tape. To install adhesive-backed tape, press it firmly against the joint. You *do not* need to apply an embedding coat of joint compound first. Run a taping knife or trowel along the entire length of the tape to bond it tightly. Use a corner tool to press the tape into inside corners. Don't overlap one piece of tape onto another. The thickness of the tape makes the overlap hard to conceal. Cover the tape with successive layers of joint compound. When done right, joint compound will conceal the tape.

Fiberglass mesh tape is also available without adhesive backing. It costs a little less than the adhesive-backed tape. Use staples to install nonadhesive tape. Fasten two staples at one end of the tape. Lay the tape along the joint and fasten two more staples at the other end of the tape. Place additional staples every 24 inches along the tape strip. Stagger these staples so successive staples are on opposite sides of the joint.

When installing fiberglass tape where walls meet ceilings, staple tape to the ceiling only. On inside corner joints between walls, you'll normally staple only one edge of the tape. But you can add staples to the other side of the tape if necessary to make it lay flat. Figure 4-2 shows nonadhesive fiberglass tape stapled into an inside corner.

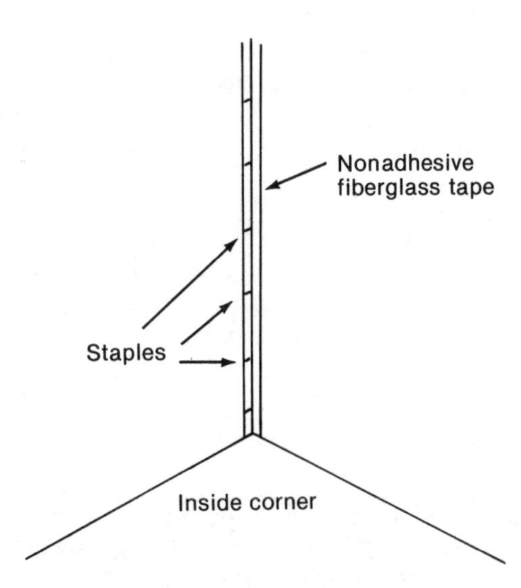

Nonadhesive fiberglass tape on inside corner
Figure 4-2

TOOLS

In Chapter 2, we discussed all the common drywall tools. Let's review the three tools you'll use most often for finishing joints: the mechanical taping tool, the drywall knife, and the compound dispenser.

You'll use the mechanical taping tool to apply the tape and compound in one operation for the embedding coat. You can use this tool on either flat joints or inside corner joints. As you roll the tool along the joint, it lays the tape in the joint and dispenses the proper amount of compound at the same time. When using this tool on vertical joints, start at the bottom and work toward the top.

After dispensing the tape and compound for the embedding coat, use a wide drywall knife to smooth out the joint and remove any excess compound. Hold the knife at about a 45-degree angle to the surface of the drywall and press firmly. You want to force excess joint compound out from under the tape. Start at the center of the tape and work toward the edges. This reduces the chance of wrinkling the tape.

Joint compound will shrink a little as it dries. The second coat of joint compound will fill the joint and even out the joint surface. Use a compound dispenser to apply the second and third coats of joint compound. This tool dispenses the compound through a slot, and feathers the edges. The second coat should be about 2 inches wider than the embedding coat. Use a wider dispensing tool to apply the third coat. Each successive coat should be wider than the previous one. This conceals the joint better than if each coat were the same width.

There's a special compound dispenser made just for filling fastener depressions. This tool dispenses compound and wipes off the excess compound in one operation.

Use a loading pump to load compound into your mechanical taping tool or compound dispenser. The compound you load into these tools should have a little thinner consistency than the compound you'd use if you were applying it by hand. You can thin ready-mix joint compound by adding water to it. Be sure to mix the water in thoroughly so you have a smooth consistency. There shouldn't be any lumps.

SEVEN KEY JOINT TREATMENTS

The long edge of a drywall panel is normally tapered. So where two panels meet, there's not only a joint, there's a depression. See Figure 4-3. In this depression you apply joint tape and compound. When finished, the area over the joint should be flush with the rest of the panel. Sometimes you'll find a joint where only one of the two panels is tapered or where there's no taper at all. This makes it harder to conceal the joint. This section will explain what's required to make good joints, even at cut ends.

You should know how to make seven types of joints: angular-tapered joints, round-tapered joints, nontapered joints, inside corner joints, outside corner joints, expansion joints and joints in decorated panels.

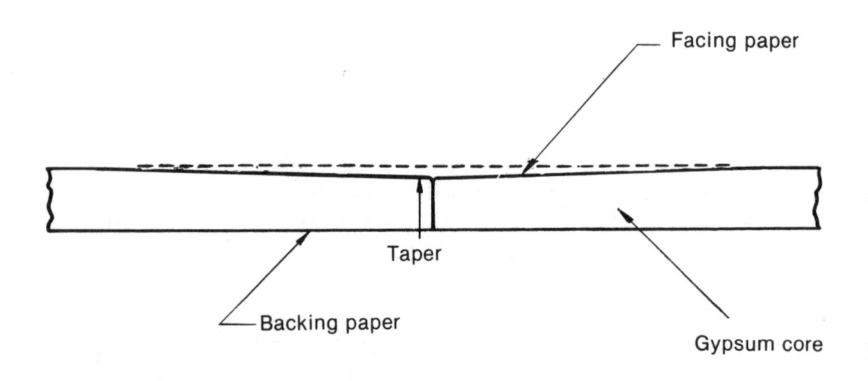

Tapered wallboard joint
Figure 4-3

Angular-Tapered Joints

The tapered edge of a drywall panel can be either an angular edge or a rounded edge. Figure 4-4 shows both kinds. Let's begin with an angular-tapered joint. You'll need joint compound, paper tape and several drywall knives. Here are the steps to follow:

1) Using a 3- or 4-inch wide knife, apply a thin coat of joint compound to the joint. Hold the knife at about a 45-degree

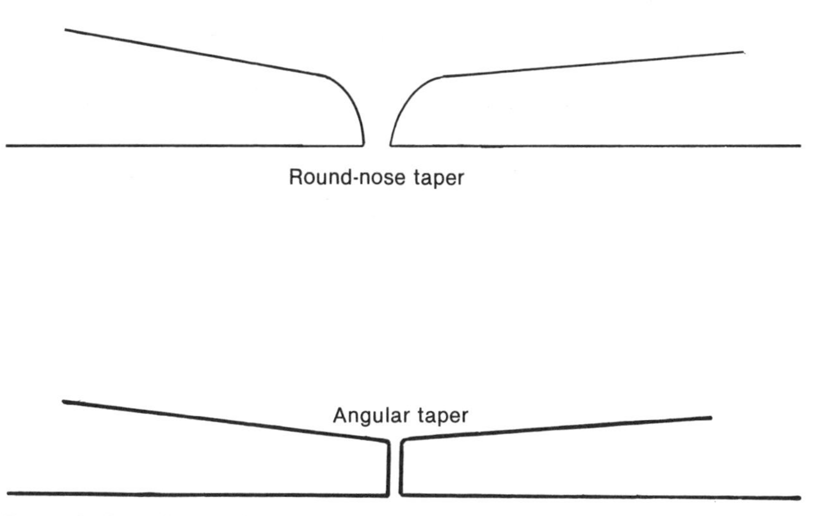

Rounded and angular panel edges
Figure 4-4

angle to the panel, distributing the compound evenly. This coat of joint compound is known as the *embedding coat*.

2) Use the same 4-inch knife to press joint tape into the embedding coat. You can apply dry paper tape, but many finishers prefer to moisten it before application. To premoisten the paper tape, just run water over it briefly and shake off the excess water. Dry tape draws moisture out of the joint compound. If it draws out too much moisture, the bond between the tape and the compound will weaken.

3) Immediately after you press the tape into the embedding compound, use the knife to apply a very thin coat of joint compound over the tape. This *skim coat* of compound helps keep the tape from curling or wrinkling. Tape exposed to the air loses moisture faster than the rest of the joint.

A bubble will form in joint tape anywhere tape isn't properly bonded to drywall. This bubble makes it harder to conceal the joint. If you see a bubble form, slit the bubble with a knife and work some joint compound under the edges. Then press it flat. Sand between coats, as necessary. And be careful not to scuff the tape.

4) While the skim coat dries, apply a coat of joint compound to the fastener head depressions. This is known as *spotting* the fastener heads. Fill depressions so they're flush with the surface of the panel. This is also the time to apply joint compound to all corner beads. Use the panel surface and the bead to guide the knife as you apply compound. Feather compound so it's thinner at the edge. That conceals the transition between compound and the panel surface.

5) Allow the embedding coat, skim coat and fastener-head spotting to dry thoroughly. This may take 24 hours or longer, depending on the temperature and humidity of the air. Now you can apply the next coat of joint compound. If you're using a fast-drying joint compound, be careful not to mix too much at one time. It can harden very quickly.

Use a wider knife to apply the second coat of joint compound. Feather the second coat out about 2 inches beyond the first coat. This helps the second coat blend into the panel, making a joint that's easier to conceal. Then apply a second coat to the fastener heads and to the corner bead, feathering each into the panel surface.

6) After the second coat dries, look it over carefully for smoothness. In dim light you won't see all surface flaws. Run your hand over the surface, looking for depressions that need more compound. Then sand lightly, if needed, to smooth the second coat. It's important that this coat be very smooth.

Any raised sections or bumps in the second coat will get worse when the third coat is applied. Here's why. Suppose your knife hits a bump when applying the third coat. What happens? The knife raises up, moving slightly away from the surface. This deposits more compound in the area around the bump. When that compound dries, the bump has grown to a ridge in the new coat. This ridge will be parallel to the knife blade and will take even longer to sand smooth.

Be careful to sand lightly. Don't scuff the joint tape. Scuffing causes little balls of paper to form on the tape surface. This makes it harder to get a smooth finish in the next coat of compound.

There are two common sanding methods: dry sanding and wet sanding.

Dry sanding— Use open-coat sanding paper so that it doesn't clog up quickly with sanding dust. The 100-grit paper is about the coarsest grit you can use and still get a smooth surface. Be sure to wear eye protection when sanding. A breathing mask is particularly important when you dry sand.

Wet sanding— Wet sanding creates less dust than dry sanding. You can use wet or dry sandpaper and water to wet sand. Or you can use a synthetic-fiber abrasive pad with water. If you wet sand, the compound will absorb some of the moisture and will have to dry out again. But it won't take as long as when you first applied the compound.

7) It's important to remove all of the sanding dust before applying the third (or finish) coat of joint compound. Joint compound won't stick over sanding dust. Any sanding dust on the surface will cause grooves in the compound as you draw the knife along the joint.

When the surface is smooth and free of dust, apply the third coat of joint compound to the taped joint, the fastener heads (if necessary), and the corner bead. Feather the third coat out about 2 inches beyond the second coat.

Using a mechanical taping tool— The procedure is about the same when you're using a mechanical taping tool. Follow the manufacturer's instructions for loading, using and cleaning the tool after the job is done. The taping tool will apply both the embedding coat of joint compound and the paper joint tape at the same time.

As soon as you've applied the tape and embedding coat of compound, go back over the joints with a broad taping knife to smooth the joints. This forces excess joint compound from under the tape and leaves a skim coat of compound over the tape. You can use a corner tool to wipe down the corners, leaving a smooth surface.

After this first coat of compound has dried thoroughly, apply the second coat of compound to the joints. You can use a corner applicator and finishing tool for applying this coat to achieve a smooth finish.

Using a compound dispenser— You can save even more time by using a compound dispenser to apply compound to the fastener heads. All of these tools hold and dispense a measured quantity of compound when they're moved over the surface of the drywall.

You can cover territory much faster using these mechanical tools than you can with a taping knife. If you can't avail yourself of all these tools, just using some of them will still save time. For instance, if you use an mechanical taping tool or a banjo to install tape and the embedding coat of compound and do the rest of the finishing with a taping knife, you'll have gained a considerable amount of time over doing the whole job manually.

Round-Tapered Joints

Finishing round-tapered joints is only slightly different.

1) Fill the "V" where the two rounded edges come together. Use a 3- or 4-inch wide knife to apply the joint compound. Then remove all excess joint compound so the surface will be smooth.

2) Let the joint compound harden. While you're waiting for it to harden, apply the first coat of compound to the fastener heads.

3) Apply the first coat, tape, second coat and third coat, the same as for angular-tapered joints.

Nontapered Joints

Two common nontapered joints are panel end joints and joints between panels that have been cut and are no longer tapered at the edges. Nontapered joints are more difficult to conceal than tapered joints.

Where there's no taper, there's no depression along the joint. When you apply joint compound and tape to a nontapered joint, it creates a raised section along the joint. This is also true where you have a tapered edge of one panel butting up against a nontapered edge of another panel.

You can still make a good joint over nontapered edges by feathering the joint compound out farther than you did for tapered joints. The width of the final coat of joint compound may have to be as much as 18 inches. And it's important to apply enough joint compound under the tape to bond the two panels securely.

Inside Corner Joints

Finishing inside corner joints takes more care and time than finishing other joints. It's hard to do anything to one side of the joint without affecting the other side.

For instance, when you draw a taping knife down one side of the joint to apply or smooth joint compound, the edge of the knife will probably rub along the wall surface on the other side of the joint. If the compound on the other side is wet, the knife will make a groove in it. You can overcome this with a very steady hand, or by waiting for one side to dry before doing the other side, or by using a corner-finishing tool.

The most common corner-finishing tool is a taping knife that has a flexible blade bent at slightly more than 90 degrees. With this tool, you can smooth both sides of the joint at once by drawing the tool down the center of the joint. Since the angle of the blade is slightly *more* than the angle of the corner, the blade springs against both sides of the corner as you draw the knife down the joint. This lets you smooth the entire corner at one time. The tool also feathers the edges of the compound. The knife applies more pressure at its outside edges than it does at the center of the blade.

Another type of corner-finishing tool is hinged at the center, rather than bent. This tool has a spring that forces the two halves apart. The distance they are forced apart is limited by stops. So the angle formed by the two halves is slightly more than 90 degrees. Use this tool the same way you'd use the bent-blade corner-finishing tool.

To tape and finish an inside corner, follow these steps:

1) Apply an embedding coat of joint compound to both sides of the corner.

2) Fold the joint tape along the center crease. Lay it in the corner and press it into the joint compound with a corner-finishing tool. Press any excess compound out from under the tape.

3) Spread a thin coat of joint compound on top of the tape. Smooth the compound, feathering it out to the edges. Figure 4-5 shows a wall-ceiling joint after the first three steps are complete.

4) When the joint compound is completely dry, sand it lightly with open-coat sandpaper, 100-grit or finer. (The grit number

increases as the paper becomes finer.) Be sure to use a breathing mask and eye protection while sanding. Remove all sanding dust.

5) Apply a second coat of joint compound, feathering it beyond the edges of the first coat.

6) When the second coat is dry, sand it lightly, remove the dust and apply the final coat of joint compound. Feather it about 2 inches beyond the edges of the second coat.

Wall-ceiling joints with tape applied
Figure 4-5

Outside Corner Joints

There are two common ways to finish outside corner joints. One method uses corner bead. The other uses tape.

Corner bead— In most cases, you'll use corner bead on outside corner joints. Corner bead is made of metal, generally galvanized steel. It normally serves three purposes: protecting the corner from damage, providing a screeding surface for the taping knife, and holding the two panels together.

Corner bead comes with solid metal flanges or metal mesh flanges. Use a special clinching tool or drywall nails to install bead with solid metal flanges. Use staples to install the bead with mesh flanges. Space the nails or staples 9 inches apart on both types of bead.

Install corner bead in one continuous piece, unless the corner is too long to do this. It's hard to get a perfect match between two pieces of corner bead. Installing more than one piece at a corner makes it harder to get a smooth finish. Figure 4-6 shows a corner bead installed around a square opening.

Corner bead around square opening
Figure 4-6

Here's how to install corner bead on an exterior corner:

1) Apply the first coat of joint compound with a taping knife that's wide enough to apply the compound 1 or 2 inches beyond

the flange of the bead. Use the bead as a screed (or guide) for the blade of the knife.

2) After the first coat dries, sand where necessary.

3) Apply a second coat, feathering it about 2 inches beyond the edge of the first coat. Follow the same procedure for the third coat.

When installing corner bead on a curved surface, such as around an archway, use flexible beading. Make cuts across the flange of the bead so you can bend it to fit the curve of the arch. The cut flange goes against the flat wall adjacent to the arch. Figure 4-7 shows a corner bead installed around an archway.

Taping outside corners— Some outside corners don't need corner bead. Outside corners at soffits and above cabinets, for example, don't need corner bead. The corner treatment only has to blend in with other wall or ceiling joints. Any exterior corner at eye level or higher doesn't need bead for physical protection.

Corner bead around archway
Figure 4-7

Use a flat taping knife or outside corner-finishing tool to apply joint compound to these outside corners. The outside corner-finishing tool is a taping knife with a flexible blade bent in the middle. The bend is slightly less than 90 degrees. This lets the blade press against both surfaces of an outside corner at the same time, and simplifies embedding tape and feathering edges. It also gives you a smooth and slightly rounded corner.

Where two pieces of gypsum drywall meet at an outside corner, the edge of one of the pieces will be exposed, as shown in Figure 4-8. You'll need to finish this corner so that it's smooth. You can use either corner bead or joint tape to finish the corner. Joint tape will bond the pieces of drywall together. It will also strengthen the joint. And it provides a smooth surface for the second and third coats of joint compound.

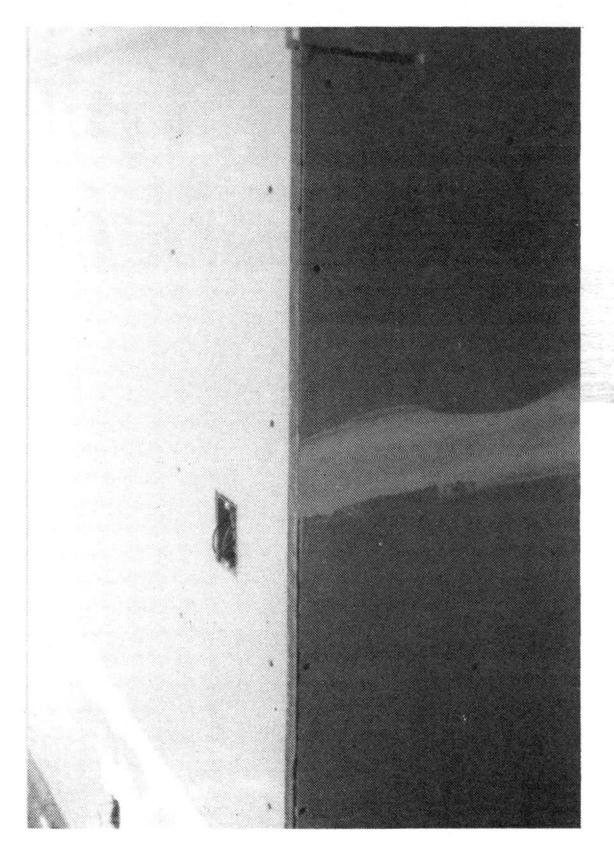

Unfinished corner
Figure 4-8

To apply joint tape to an outside corner, follow these five steps:

1) Apply an embedding coat of joint compound to both of the surfaces that form the outside corner.

2) Fold the joint tape along the center crease. If you're using joint tape around an archway, crease the tape down the center and make cuts halfway across the tape, from one edge to the middle. This lets you bend the uncut section of tape around the inside of the arch. The cut half of the tape will lay flat against the wall that's next to the arch.

3) Lay the tape along the joint. Using a taping knife or an outside corner tool, embed tape in the compound. Force the excess joint compound out from under the tape.

4) When the embedding coat is dry, sand lightly so the next coat goes on a smooth surface.

5) Apply a second coat of joint compound, feathering it out about 2 inches beyond the first coat on both sides of the joint. When the second coat is dry, apply the final coat the same way. Feather it beyond the edges of the second coat by about 2 inches.

Expansion Joints

All building materials expand and contract with temperature changes. Although the amount of expansion or contraction is small, over a large area it can be enough to buckle drywall at the weakest point. Expansion joints help avoid this problem.

Expansion joints provide *controlled* changes at the weak point. Limited movement won't cause damage that's obvious. Figure 4-9 shows the most common situations that require expansion joints.

Let's take a look at this figure. It shows that an interior ceiling *with* perimeter relief should have an expansion joint every 50 feet or less. It also shows that an interior ceiling *without* perimeter relief should have an expansion joint at least every 30 feet. *Perimeter relief* means that the edges, or perimeter, of the ceiling are not restrained (or are relieved) from expanding. The permissible distance between expansion joints is greater where a ceiling has perimeter relief, because the ceiling is free to expand around the edges. That means there's less tension locked up in the ceiling than when there's no perimeter relief.

Wallboard installation	Maximum span	Maximum area
Partitions	30'	---
Interior ceiling with perimeter relief	50'	2 ,500 SF
Interior ceiling without perimeter relief	30'	900 SF
Exterior ceiling	30'	900 SF

Expansion joint applications
Figure 4-9

There are two common ways of installing expansion joints. One is to attach an expansion joint to the drywall panel. The other is to attach an expansion joint directly to the joists before hanging board. Let's look at both of these methods.

Attached to drywall panels— This method is shown in Figure 4-10. Follow these steps:

1) Leave a 1/2-inch gap between the drywall panels.

2) Use a fine-tooth hacksaw to cut the expansion joint to the proper length. If you need to butt two pieces together to make a complete joint, cut the ends so they fit together neatly.

3) Fasten the expansion joint to the drywall with 9/16-inch staples spaced at 6 inches along each flange of the joint. The flanges have holes closely spaced so you can place the fasteners anywhere along their length.

4) Finish the joint with joint compound.

5) Remove the strip of plastic tape that runs down the center of the expansion joint. The plastic tape helps you finish the flanges smoothly. Removing the tape leaves the joint open.

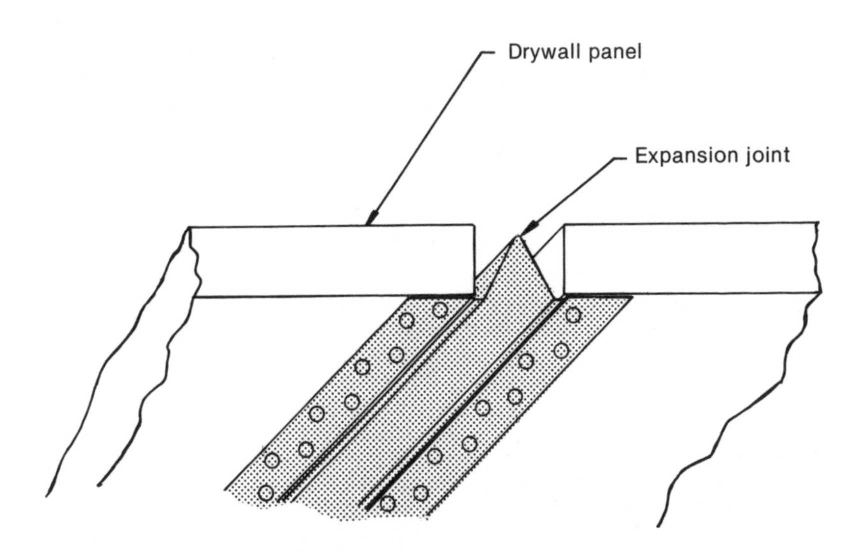

Expansion joint attached to drywall
Figure 4-10

Attached to joists— This method is commonly used in radiant heat ceiling installations. See Figure 4-11. Here are the key points to remember when attaching an expansion joint to ceiling joists:

1) Joists must be close together at expansion-joint locations to permit fastening the joint's top flanges to both joists.

2) Where two ends of an expansion joint butt together, splice the ends with 16-gauge galvanized wire.

3) Fasten the expansion joint to the joists with 9/16-inch staples (or nails) spaced 6 inches along each flange of the joint.

4) Finish the joint with joint compound.

5) Remove the strip of plastic tape that runs down the center of the expansion joint.

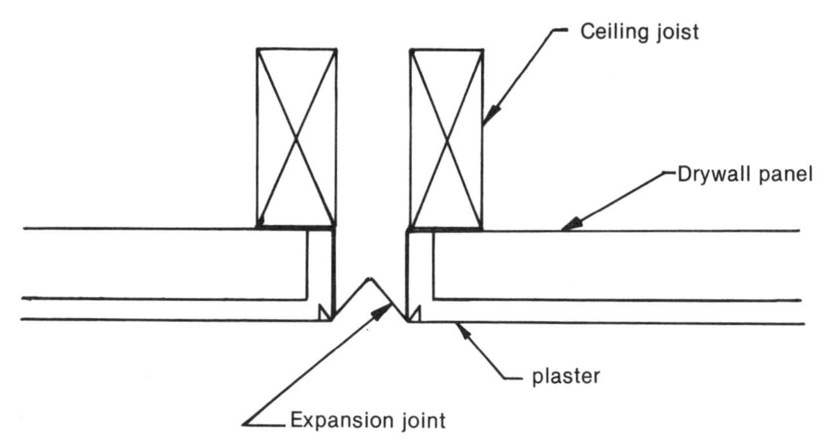

Expansion joint attached to joists
Figure 4-11

Joints in Decorated Panels

Decorated panels arrive from the factory already finished with various coverings. Joints for these panels aren't covered with compound and tape. Instead, some panels come with a flap of covering material that spans and conceals the joint. On some panels, edges are wrapped in covering material so the joint shows a finished seam. The panels can also be designed to accept battens or trim that snaps into place between the panels.

Vinyl coated drywall is popular in commercial buildings because it never needs painting and is cleaned easily with soap and water. It's seldom used in residences because any unevenness in the wood frame behind the board will be more evident than if joints are taped and sanded. But over metal studs, vinyl coated drywall may be a good choice, especially in offices and stores.

Chapter 5

SURFACE TREATMENTS

The first drywall was made for patching holes in plaster walls. A small section of drywall was cut to fit over the hole. Then the patch was textured to match the surrounding plaster, usually by troweling plaster over the patch.

Drywall gradually replaced plaster as an interior wall cover. But because builders wanted to retain the look and feel of plaster, they began applying surface treatments to drywall to give it that appearance. Now many types of drywall surface treatments are available. Many still imitate the look and feel of a plastered wall.

The one type of drywall panel that doesn't require surface finishing is the factory-decorated panel. These panels come already covered with paper or vinyl that is colored, patterned or textured. Once you hang the panels and finish the joints, the installation is complete.

The specs on most of your drywall jobs will require that you apply some type of finish to the panel surface. This chapter will explain all the popular finishes, from smooth to heavily textured, including acoustic ceiling finishes. We'll begin with a look at some basic finishing principles. Then we'll discuss the materials, tools and application techniques for five types of surface treatments: trowel finish, skip-texturing, texture paint, acoustical ceiling coatings and veneer plastering. Finally, I'll explain how to refinish a textured wall.

SELECTING THE RIGHT SURFACE TREATMENT

When selecting the material and method for surface finishing, be sure to consider the end use of the room itself. Some rooms need a smooth finish; others look best with a rough-textured finish.

You'll want a smooth finish in rooms with walls that require frequent cleaning, such as kitchens, bathrooms and children's playrooms. Rooms that have a lot of airborne dirt should also have smooth-finished walls. Dirt particles are less likely to stick to a wall with a smooth finish. But keep in mind that the labor hours required to do a smooth finish are greater than the time required to do a rough finish. A perfectly smooth finish requires more time because even a slight surface defect or imperfection will be obvious in the finished surface.

A rough- or thick-textured finish conceals surface imperfections. And the variety of textures you can produce is limited only by your imagination. Some texturing compounds can be sprayed on. Others are applied with a brush or roller. And you can vary the end result by blotting, rolling, brushing, sponging or troweling the surface before it dries.

Basic Guidelines

Common finishing problems include discoloration, odors and breakdown of the texture. The usual cause is too much bacteria in the texturing tanks. Here's how to avoid this problem:

- The moisture in the compound aids bacterial growth. Don't mix more compound than you're going to use in a 24-hour period.

- Don't expose the mix to high temperatures or direct sunlight for long periods of time. High temperatures promote bacteria growth.

- Clean your texture tanks or other containers at least once a week. Use a mixture of chlorine bleach and water — one gallon of bleach to one hundred gallons of water. Cycle the mixture through the tanks and hoses. Then let it sit overnight.

- Use only clean tools and potable water for mixing.

With those rules in mind, let's look at five types of surface treatments. For each treatment, we'll discuss the surface effect, the materials and tools required, and the application technique. Then we'll look at how to refinish a textured wall.

TROWEL FINISH

When you want the finished surface to look like a plaster wall, use a *trowel finish.* There are several variations to choose from. You can vary the depth of the texture coat and the amount of troweling done. You can use a very light texturing compound or a very heavy one.

Materials

For texturing, use standard joint compound, all-purpose joint compound, topping compound or a premixed texturing compound. Or you can create your own compound material by adding a texturing agent to a joint compound. There are a number of texturing agents you can add. There are commercially available agents in different grades of coarseness, or you can be inventive. Add dry compound powder to the compound mixture for a fine texture. You can add clean sand for a coarser texture. The coarseness of the texture will also depend on how much of the texturing agent you add in proportion to the compound. Experiment with small amounts. As always when you experiment, keep track of what and how much you add so you can repeat a successful combination.

Texturing agents tend to dry the compound. This causes it to "tear" as you trowel it onto the surface of the drywall. Let's look at an example. Assume you're using a trowel to apply plain joint compound to a drywall surface. You'll apply pressure as you draw the tool across the surface. If you're holding the tool so that the trailing edge presses hardest on the surface, you'll leave a smooth coating of compound. If you add a texturing agent to the compound and apply the same kind of pressure, you'll get some tearing in the surface instead of a smooth coat. This gives the finished job a roughened texture, which may or may not be desirable.

If you trowel the joint compound too much, you'll get the same sort of tearing, even when there's no texturing agent in the compound. This is because repeated troweling dries the compound. This drying happens even faster when you overwork topping compound. Besides the tearing effect, texture compound will add body and roughness to the finished texture.

Tools and Technique

You can apply compound with either a trowel or a taping knife. The longer the trowel or the wider the knife, the more strength and stamina you'll need for application. You can cover more surface area using the larger tools, but you may get tired sooner. Larger tools can also be harder to control. Experiment until you find the tool size that gives you the best compromise between ease of use, control, and high productivity.

Hold the trowel or knife so there's nearly even pressure across the blade, from the leading edge to the trailing edge. This lets you spread an even thickness of compound over the surface of the drywall. The surface should have the appearance of plaster, as shown in Figure 5-1. Figure 5-2 shows the technique.

You can vary the surface pattern by changing the direction of your trowel strokes. Circular strokes will give you a swirled pattern. Make a straight pattern with straight strokes.

Here are some suggestions for applying a trowel finish:

1) Select a trowel that's between 10 and 24 inches long. The shorter the trowel, the easier it is to control.

2) Using a putty or taping knife, build up some compound on the blade of the trowel. Some finishers prefer to use a wide pan and scoop up compound with the trowel. The exact amount of compound isn't critical, as long as it extends at least the length of the trowel blade. Start with an amount that's about 1/2 inch wide by 1/2 inch thick and runs the full length of the blade.

3) Hold the blade so the trailing edge is nearly touching the drywall surface. The leading edge should be raised about an

Trowel finish
Figure 5-1

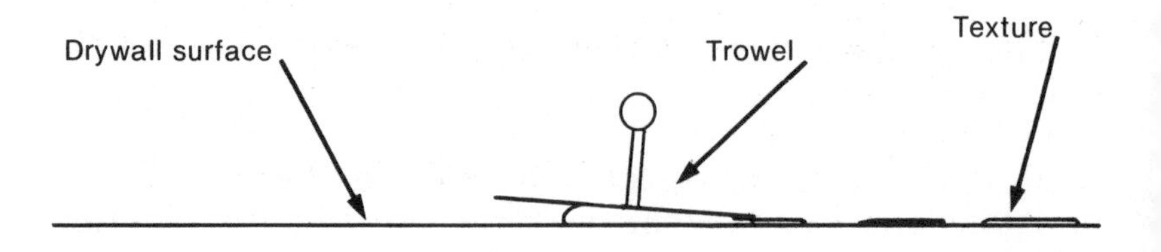

Skip-troweling
Figure 5-2

inch off the surface. Sweep the trowel across the surface of the drywall, applying compound about 1/16'' deep.

4) When you've spread all the compound on the blade, trowel back over the surface again, evening the layer of compound and finishing the texture. Vary the angle of the trowel blade and the pressure to get more or less tearing in the surface.

5) Repeat steps 1 through 4 above until the entire drywall surface is covered.

You can then vary the final surface texture by treating the surface before the texture compound dries. Here are two variations you might want to try:

- Apply the texture compound and then blot it with crumpled newspaper, foil, a brush or a sponge. Blotting the texture is time-consuming. If this is what the customer wants, be sure to include the additional labor in your estimate.

- Once you've blotted the surface, try varying the effect by flattening the raised portions with a knife or trowel. Figure 5-3A shows a blotted finish that's been lightly swept with a trowel. Figure 5-3B shows a blotted finish swept with more pressure applied to the trowel.

SKIP-TROWELING

When you want a surface treatment that's similar to a trowel finish but with a more random texture, use *skip-troweling*. Skip-troweling is also known as skip-texturing, brocade finish, and perhaps many other names.

This texturing method is easy to do and quite fast. Where a trowel finish spreads texture compound over most of the panel surface, skip-troweling, as the name implies, spreads it in blotches. You can control the size and height of these blotches to control the roughness of the finish. The panel surface that has no texture compound applied (the area between the blotches) will be perfectly smooth. Skip-troweling uses less texture compound, as the compound may cover only about 50 percent of the panel surface.

Blotted finish lightly swept with trowel
Figure 5-3A

Blotted finish swept with more pressure on trowel
Figure 5-3B

Materials

When applying joint or topping compound to finish a joint, you're after a smooth finish. The smoother and creamier the compound, the smoother the final finish will be. When applying texturing compound for skip-troweling, you're after a rougher, more random finish. Adding drying agents, such as sand, to the compound will increase the roughness of the finish. No. 30 mesh sand is normally a good drying agent. Experiment with the amount of sand you add until the finish is just right.

Tools and Technique

Use a taping knife or trowel for skip-troweling. The wider the knife or trowel, the more area you can cover with each stroke. Here are the steps to follow for one method of skip-troweling.

1) Dip the edge of the knife blade or trowel into the texturing compound. There should be compound along the full length of the edge of the blade.

2) Place the edge of the blade so it's nearly touching the drywall surface. The compound built up on the edge of the blade will be touching the surface.

3) Holding the blade at slightly less than 90 degrees to the surface of the drywall, draw the blade across the surface. Decrease the angle of the blade as you draw it across the surface so all of the compound is applied to the surface. Use a light touch. Too much pressure will apply compound too fast. The idea is for the compound to be applied in globs covering about half of the drywall surface. These globs should be irregular in shape and location.

4) After you've applied all of the compound you had on your knife, you may want to draw the knife lightly across the compound to flatten any high spots.

5) Repeat this process until you've covered all of the areas you want covered. Remember that you'll be leaving some of the drywall surface uncovered. See Figure 5-4.

You'll see finishers do skip-troweling by applying random blobs of compound to the drywall, either with a sprayer or by

Skip-troweling results
Figure 5-4

dabbing it on with a taping knife. Then the finisher goes back to flatten the surface with a trowel or knife. My experience is that dabbing it on with a knife takes a little longer, while spray application can be faster. But you may find the added control of knife application an advantage, especially if you want to apply texture on a limited area such as a patch where you're trying to match the rest of the surface.

TEXTURE PAINT

When you want a lightly textured finish that's easy to clean, use *texture paint*. It consists of a heavy latex paint with a texturing additive. It's slightly thinner than joint compound but much thicker than regular paint. When applied with a roller or sponge, texture paint gives you a *stipple* finish that will hide minor imperfections and still leave a relatively smooth finished surface. See Figure 5-5.

Stipple finish
Figure 5-5

Materials

Texture paint comes in a ready-mix formula or in a dry compound you can mix yourself. Premixed texture paint is expensive but more convenient to use. If you decide to mix your own compound, remember these important points:

- You can increase the bonding power and the hardness of the texturing compound by adding latex emulsion.

- Careful preparation is important. Read the manufacturer's directions carefully. Don't overthin the mixture. Overthinning reduces the amount of texture in the compound. It also weakens the adhesive in the mixture so it won't stick as well to the drywall.

Tools and Technique

Texture paint is easy to apply. Use a roller, a sponge or a brush. A roller is probably the fastest applicator. The final

Stipple overlay on skip-texture finish
Figure 5-6

surface texture will vary, depending on the type and thickness of the roller nap. The longer the roller nap, the higher and farther apart the stipple peaks. Apply a thick, even coat.

Figure 5-6 shows a skip-texture finish covered with texture paint to provide a stipple finish overlay. A roller was used for the stipple finish.

ACOUSTICAL CEILING COATINGS

Acoustical ceiling coatings are used frequently in new construction and are often added to older homes. A rough acoustical ceiling contrasts nicely with smooth walls and will hide most surface defects in the ceiling. Joint and fastener finishing don't have to be perfect when the ceiling is going to be coated with an acoustical finish.

A ceiling finish (or any finish) isn't acoustical unless it *absorbs* sound, not merely reflects it. Drywall isn't a good sound-absorbing material. It reflects more sound waves than it absorbs. If you're using an acoustical ceiling coating as a decorative finish only, it's O.K. to apply it directly to drywall panels. But if you want the coating to function as a sound-absorbing finish, you'll need to install a sound-absorbing base material *before* you apply the acoustical coating. The coating itself does little to absorb sound waves.

Basic Materials

Fiberboard is a good sound-absorbing material. It's a lightweight wood-fiber product available in 4-foot by 8-foot sheets that are 1/2 inch thick. It also comes in panels with smaller dimensions. Wood fibers in the panels are compressed and held together by a binder or glue. Some of the panels come with 1/8-inch predrilled holes spaced over the entire panel surface.

You can fasten fiberboard panels directly to the framing members. Fiberboard panels can be applied directly to framing or on top of drywall. You can install them in suspended ceiling runners. Or you can attach them to resilient channels. When the fiberboard panels are in place, you're ready to apply the acoustical coating.

Perlite or vermiculite is normally the aggregate in the mix. The size of the aggregate determines the roughness of the finished texture. These finishes dry to a bright white color and don't need to be painted. Your drywall dealer probably sells several mixes and aggregate compounds. Select the combination that yields the texture you want.

Tools and Technique

Acoustical coatings, whether functional or just decorative, are sprayed on. The finished product looks about the same for both the decorative and functional acoustical ceiling coatings.

The first step in spraying is to mask or cover all surfaces you don't want coated with spray. One of the most common problems in applying acoustical texture is the overspray. Cover everything you don't want coated. Masking adjacent surfaces may take more time than applying the texture. But it's time well spent. You may do a perfect job of applying texture to the ceiling, but if there's overspray on other surfaces, your customer won't be happy. And neither will you be happy when you have to spend hours scraping it off.

The spraying itself is relatively easy. Just keep the coating thickness even over the entire surface.

VENEER PLASTERING

When you want a hard, durable finish for high-wear areas, such as in stairwells, use *veneer plaster*. Veneer plastering is a thin (3/32-inch) coat of plaster that you apply to *veneer-base* drywall panels. The surface of veneer-base drywall is designed to provide a good grip for veneer plaster.

Materials

Veneer plaster is available in one- and two-coat applications. Either way, the mix is a dry powder. You add water at the job site. Veneer plaster sets up quickly, so mix only as much as you can apply in a few minutes.

Follow the manufacturer's instructions carefully when adding water or aggregate to the mix. Otherwise, the mix may not spread properly or set up as it should. Mix the material just long enough to get a smooth consistency. Overmixing speeds up the setting time.

It's important to clean your equipment thoroughly before the plaster has a chance to dry on it. Veneer plaster sets up very hard and bonds tightly, especially on surfaces where you don't want it.

Tools and Technique

Veneer-base drywall requires special preparation before you can apply the compound. First you'll need to reinforce the panels at the joints. The veneer coating is too thin to add much strength where panels meet. Use corner beading on outside corners. Apply fiberglass tape to all other joints, including the inside corners.

You can use paper joint tape instead of fiberglass tape, *if* the paper tape is embedded in joint compound. Note that not all joint compounds are compatible with veneer plaster. If you're using paper joint tape, be sure to choose a joint compound that's compatible. Embed paper tape in the joint compound just as you would with regular drywall. You won't need to apply second or third coats of joint compound. One coat is enough. The veneer plaster will hide the joints. But let the joint compound dry before you apply the veneer plaster.

Apply veneer plaster either by hand or with a spray rig, using either the one- or two-coat method of application.

Manual application— When applying the veneer plaster by hand, here are the important points to remember.

One-coat method— Apply the plaster to the fiberglass-taped joints first. This keeps the tape from wrinkling as you apply plaster to the rest of the drywall surface. Using a trowel, hit the joints with a thin skim coat of plaster, about 1/16 inch thick. When this skim coat is firm, trowel on a veneer coat over the whole surface. The final thickness should be about 3/32 of an inch.

Most applicators use a brush, trowel or sponge to texture the plaster. If you want a smooth coat, use a trowel. It's easiest to get a smooth-trowel finish if you do the final troweling as the plaster starts to darken slightly. Darkening means that the plaster is starting to set. Keep a little water on the trowel to aid smoothing. But don't use too much water.

Be careful not to overwork the plaster or to work through the plaster to the base. If you want a deeper texture or a coarser finish, you can add sand to the plaster. But follow the manufacturer's instructions when adding anything to veneer plaster mix. Another choice is to use a plaster that's formulated for texturing.

Two-coat method— Two-coat veneer plaster provides an even stronger finish. It's also a good choice where you're applying plaster over a masonry base. If the masonry base is concrete block, you can apply plaster directly to it. If the base is smooth concrete, such as a poured wall, you'll have to apply a bonding agent to the concrete first. Let the bonding agent dry before you apply veneer plaster.

The base coat of the two-coat method is stronger than normal plaster. Apply the base coat and let it dry for about two hours. But leave the surface slightly rough so it can grip the finish coat. Then apply the finish coat evenly.

Texturing the final coat in a two-coat job is the same as texturing one-coat veneer plaster. Another choice is to simply trowel the surface smooth.

Spray application— You can also spray on veneer plaster. Special plasters are available for spraying. Here are some hints if you're using a spray rig to apply veneer plaster.

- Keep your spraying tools clean.
- Practice spraying on scrap material before you try working on a job. Experiment with changes in the nozzle opening, air pressure and thickness of the mix to get the result you want.
- Spray only as large an area as you can cover in about half the indicated setting time of the plaster. Then clean your equipment.
- Apply the plaster in two passes. Make the second pass at right angles to the first pass.
- If you want a smooth finish, trowel the plaster smooth after you spray it on. You may want to flatten high spots with a trowel or wide knife for a different effect.

REFINISHING A TEXTURED WALL

On a remodeling job, you may have a situation where the owner wants to apply wallpaper over a heavily textured wall. There are several ways you can handle this. You can sand the texture flat (or at least flatter). But this is time-consuming, messy work. You have to sand evenly, using a sanding block to keep the surface flat. This raises a lot of dust and takes a long time. Your best bet may be to cover the existing texture with joint compound or drywall. Let's look at both of these alternatives.

Joint Compound Method

This may or may not work, depending on how rough the existing wall is. To remove the highest spots, try spot sanding. Once you've done this, use a trowel or wide taping knife to apply joint or topping compound to the wall.

If you don't use a texturing additive and you don't overwork the compound, it should go on smoothly. The compound will fill in the depressions in the old finish. The end result will be a smooth surface. It may not be possible to get a wall as smooth as on a new plaster job, but it should be close. When the compound has dried, you can smooth the surface with a moist abrasive pad as necessary.

Coating with joint compound is relatively fast. Most important, it leaves a finish that's smooth enough for any type of wallpaper or wall covering.

Wallboard Panel Method

The major disadvantage to applying new drywall on top of old drywall is that you'll have to tape and conceal all of the joints, including those at the ceiling and wall intersections. This increases your labor costs. You'll also have to match the finish on adjacent surfaces. To keep the addition from being too obvious, you may have to retexture adjoining surfaces too.

If the old surface is in good condition, apply 1/4-inch drywall over the old surface. Space the fasteners as you would in a single-ply job. You could also apply the new panels with adhesive and the number of fasteners required for a new installation using adhesive. But if the old wall surface has a very rough texture, it may not be practical to use adhesive. But using adhesive plus *extra* fasteners should be acceptable.

Tape and conceal the joints as if this were a new installation. Match the texture and finish on the adjoining surfaces. Since you're after a perfectly smooth finish on the new drywall, you may need to do some additional sanding on the joints to cover any minor imperfections.

To check for surface defects, use *oblique lighting*. This is light held at an angle to the surface. It's important to repair any surface defects that you find. Otherwise, they'll show up through the wallpaper on your finished wall.

Chapter 6

SOLVING COMMON DRYWALL PROBLEMS

Even the simplest drywall installation can have problems. The four main problems are: the fasteners, the joints, the compound and the drywall panels themselves. In this chapter, I'll take a detailed look at these four areas and suggest how to overcome the most common problems.

FASTENER PROBLEMS

Two common fastener problems are nail "pops" and fastener depressions.

Nail Pops

When drywall nail heads work up from under the surface after the installation is complete, the job is said to have *nail pops.* That's unsightly, of course. And if enough nails pop out, the drywall will loosen and sag. The nails can be driven again and the holes refinished. But the best remedy is preventing nail pops before they happen. Here are the five reasons for nail pops.

- Framing with a relatively high moisture content will shrink as it dries out. As the wood shrinks, the nails loosen.
- When drywall is fastened to framing that's out of alignment, stress on the drywall causes fasteners to work up above the surface.
- Gravity acting on ceilings and vibrations acting on walls will tend to work the nails loose.
- If a building has poor ventilation or an inadequate heating system, large temperature cycles will cause expansion and contraction of the framing and drywall. Very much of that, and the fasteners will begin to work loose.
- The drywall may not have been installed properly in the first place.

Preventing nail pops— Once you know the cause of nail pops, prevention is easier. To prevent nail pops, follow these key rules:

1) *Make sure your framing lumber is dry before you fasten any drywall to it.* The builder should provide enough ventilation to speed up the drying process. If you're working in cold or humid weather, use portable heaters or blowers to warm or circulate the air.

It may take only a few days to reduce the moisture content of lumber to an acceptable level, depending on temperature and humidity. Lumber is too green for hanging drywall if you can see a little wet spot when it's hit sharply with the head of a hammer. Test several lengths before forming an opinion. The amount of moisture can vary from one piece to the next.

2) *Make sure the framing members are aligned in the same plane.* Sight along the edges of the studs and joists to see that

they are in a straight line. You can also check alignment by holding a long straightedge up against the studs and joists.

If framing is out of alignment, repair it. Applying drywall panels to framing that's out of alignment will be a disappointment. The framing will eventually spring back to its original position. When it does, nails will pull out of the framing.

Nail pops may show up days or weeks after installation is complete, or gradually, over a period of many months. How soon they appear depends on the degree of misalignment, the type of fasteners used, the moisture content of the framing at the time of installation, the amount of vibration present and the temperature cycles that the drywall and framing are exposed to.

3) *Use ring-shank drywall nails or screws to fasten the drywall panels to the framing.* They have more gripping power than plain-shank nails and offer more protection against nail pops.

Consider using double rather than single fasteners. Double nailing gives more holding power than single nailing. You can also use floating corner angles to reduce stress on the drywall. Finally, you can use adhesive in addition to nails or screws to fasten the drywall panels.

4) *Always nail from the center of the drywall panel toward the edges.* If you nail from one edge of a panel to the other, the panel may not rest firmly against the framing for the entire length of the panel. Let's look at an example.

Assume that you're installing the final panel in a wall. The other panels along that wall are already in. And so are the panels that cover the wall that forms the other half of the corner. On this final panel, if you start installing the fasteners at one edge instead of at the center, the drywall may move slightly toward the opposite edge. This particular panel is trapped by a corner and other drywall panels. Any movement after you start nailing will stress the panel, eventually causing it to bow and pop the fasteners. A similar problem can occur if you work from the other edge toward the center of the panel. The center may actually spring away from the framing.

If you fasten the center of the panel first and work toward the edges, the panel won't be able to move once the first fasteners are installed. From the moment the first nail is driven home, the panel is forced flat against the framing in the center and along the edges.

But be sure to hold the drywall panel tightly against the framing as you install the fasteners. This ensures that the panel stays flat.

And don't worry about installing a panel that has a slight bow in it. If you install the panel properly, any stress put on the panel by flattening the bow will relax in a short time.

Before you cover the nail heads on any drywall job, check to be sure the nails are driven tight. Redrive any nail heads that are loose. It's a good idea to drive another nail on each side of a nail that's worked loose. Drive them about 1½ inches away from the old nail. After you redrive any loose nails and add extra nails, go back and check all the nail heads again. The vibration of hammering may have loosened more nails. A few seconds spent now can save hours later.

If you applied drywall to both sides of a wall, driving nails on one side of the wall may loosen nails on the other side. Check the first side again.

Attention to detail should prevent most nail pops. Take the time to check your framing lumber carefully and install your panels properly.

Repairing nail pops— When nail pops occur after you've finished the job — perhaps even after the texturing and painting are done — fixing them takes a little more time. When the nail head works out of the framing, it will show above the surface of the drywall. It may even lift the compound from the depression around the nail head. Here's how to repair it.

- First, drive the nail back in. But if the nail head has worked loose and become visible, just driving it back in may not be a long-term fix. It's better to pull the nail and replace it. The best and most permanent replacement is a drywall screw. The next best fix is a longer nail, and/or another nail within 1½ inches of the first.

- Once the fasteners are in, fill the depression with compound and let it dry. If necessary, refill it with a second coat of compound.

- You may have to repair the texture, depending on the wall's original texture. Be sure the texture of the patch matches the surrounding texture of the wall.

- When the new texture has dried, paint it with an oil-base primer. If you skip this step, the paint may soak into the repaired spot when it's repainted, making the repair stand out like a sore thumb.

Fastener Depressions

A *fastener depression* is a depressed area over the fastener head. It's the opposite of a nail pop. The joint compound over a nail or screw has sunk lower than the surface of the surrounding drywall. Fastener depressions are caused by:

- Nails dimpled too deeply or screw heads driven in too far.

- Not enough compound applied to the fastener heads to cover them properly.

- Extremely dry framing lumber absorbing moisture, squeezing the panel between the nail head and the edge of the stud or joist. In effect, the fastener head has been pulled deeper into the drywall.

- Not enough fasteners to hold the drywall firmly against the framing. In this case, the drywall may be too free. If it flexes independently of the framing, it will force the fastener heads deeper into the surface.

To prevent fastener depressions, avoid driving fasteners through the facing paper. Install the correct number of fasteners and space them properly. Spot the fastener heads with two coats of compound, sanding lightly between coats if necessary.

Repairing fastener depressions is a simple matter. First, be sure you've installed enough fasteners. If you need more nails or screws to hold the drywall firmly against the framing, add them. Second, spot the fastener heads with joint compound to bring the surface flush with the surrounding drywall.

JOINT PROBLEMS

There are seven common joint problems you'll run into in your drywall work: ridging, tape photographing, joint depressions, high joints, discoloration, tape blisters and cracks in the joint. Let's look at each of these separately.

Ridging

When a ridge occurs along a joint between two drywall panels, it's often because there has been movement at the panel joint. There are three probable causes of ridging:

- High humidity, poor heat distribution or not enough ventilation in the room. This results in expansion and contraction of the framing and the drywall panels. One cure is to install back blocking in areas where poor ventilation or heat distribution is likely to be a problem. Another choice is to install multiple layers of drywall.
- Drywall that's not properly installed. Improper installation includes misaligned framing and butt joints that don't fit together well. If you force two panels together, joint compound may be squeezed out, forming a ridge. Make sure the joint isn't overstressed by too tight a fit. You may have to cut some of the drywall away at the joint to make a little gap between panels. You can do this by cutting with a knife, making several passes. Or you can use a hand saw or a chisel to remove a sliver of one of the panels.

  If the space left between panels is too wide, the joint will be weaker than a properly fitted joint. Ideally, the entire width of the joint tape is bonded to a drywall surface. The tape itself only needs to span a small gap between the panels. If the gap between panels is too wide, less of the tape is bonded to the drywall. This can weaken the joint and promote ridging. You may have to overlap two pieces of joint tape to strengthen the joint. If the joint is very wide, it may be because one panel (or both) isn't securely fastened to the framing at the joint. Perhaps neither panel is directly over the framing member. You may have to install fasteners at an angle through the panels into the framing in order to make the joint tight against the framing.

  This raises a question. Earlier, I said to be sure to install the fasteners straight so that the edges of the heads don't stick up. Well, in this case they will. But you can fix it. Use a hammer blow to straighten the head (by bending the shank) after the fastener is tightly installed. You may also need to cut and install a narrow strip of drywall to bridge a gap that's too wide.

- Too much joint compound. To correct ridging caused by too much joint compound, first sand the ridge smooth. Then apply a finishing coat of joint compound. Hold a light at an angle to the area to make sure you've eliminated the ridge and left a smooth surface.

Tape Photographing

If you can still see the joint tape even after the wall is finished and painted, this condition is known as *tape photographing*. The tape may show through as a slightly different color than the finished wall. Or it may be the same color as the finished wall but have a higher or lower gloss to it. Photographing can also occur over fastener heads if you didn't apply enough joint or topping compound.

The usual causes of tape photographing are:

- Failure to force excess joint compound out from under the joint tape.
- High humidity delaying drying of the second and third coats of compound. This can let the tape show after the compound dries.
- If the tape absorbs too much moisture from the compound, the compound will shrink and conform to the shape of the tape. You can avoid this by prewetting the tape before installation.

To correct tape photographing, sand the tape edges to feather them into the surface of the drywall. Then cover the tape with thin coat(s) of joint or topping compound. Use thin coats so that you don't rewet the tape too much. Consider sealing the tape with a primer after sanding and before applying the final coat(s) of compound. This keeps the tape from drawing too much moisture out of the compound.

Joint Depressions

A *joint depression* is a valley that occurs in a joint. It will be most obvious when a light strikes the drywall at an angle. There are two common causes of joint depressions:

- There may not be enough joint compound over the joint. This can happen when joint compound mixture is too thin or when not enough joint compound is applied to the joint.

- The joint may be sanded too deep. Oversanding is especially common where sanding was done without a sanding block to keep the sanded area flat.

The cure for joint depressions is to add more joint or topping compound to the joint. Smooth and sand again to get a flat surface at the joint. Make sure the joint is flush with the surface of the drywall.

High Joints

A *high joint* is the opposite of a joint depression. It occurs when a wide section of joint is raised above the rest of the drywall surface. Like the joint depression, a high joint is most noticeable when light strikes it at an angle.

High joints are the result of:

- Too much compound built up underneath or on top of the joint tape.

- Improper feathering of each coat of compound. The edge of each coat must be feathered into the drywall surface. When this isn't done, high joints are the result.

Here's the procedure for repairing a high joint: First, sand the area as flush as you can without sanding into the tape. Then apply one or two final coats. Feather each coat into the drywall surface. Make each coat wider than the previous coat to conceal the joint.

Joint Discoloration

Joints may discolor or turn lighter or darker than the rest of the finished wall. There are three common reasons for joint discoloration:

- Moisture trapped inside the joint. Until a joint is sealed, it can absorb and give off water. If you seal a joint before it's dry, you'll be sealing water inside the joint. Trapped water will degrade the finish and discolor the surface. Be sure the joint is dry before you seal it.

- Painting when there's excess humidity in the air. Decrease the room humidity before painting.

- Using poor quality paints. Always use good quality paint. Cheap paints often give uneven coverage and seal unevenly. This increases discoloration.

Tape Blisters

A bubble in the surface of a joint is a *tape blister.* It can be several inches long or as small as a dime. Tape blisters occur when the bond fails between the tape and the first or embedding coat of joint compound. That happens because the joint is too wide, because the tape wasn't properly bedded in joint compound, or because the tape draws moisture too quickly from the compound. It can also happen where topping compound was used instead of joint compound to embed the tape.

Here's how to repair a tape blister:

1) First, slit the blister with a knife. If the blister is large, cut and remove the section of tape that came unbonded.

2) Sand or scrape out enough of the dried compound so you can embed a new section of tape.

3) Work joint compound underneath the tape, smoothing the slit in the old section or the new section of tape into the joint compound as you go. This embeds the blistered section.

4) Apply a skim coat of compound over the tape. When this dries, apply the finishing coats. Sand enough to get a smooth finish.

Joint Cracks

There are two common types of joint cracks: those along the edges of a joint and those that run along the center of the joint. Each has its own cause.

Edge cracks— Cracks along joint edges happen when the air temperature was high and humidity low when the joints were finished. This caused joint compound to dry too quickly and unevenly, resulting in uneven shrinkage. To slow down the drying rate, run a wet roller over the joint, or spray it with a fine water spray.

Edge cracks can also be caused by joint tape that has a thick edge or by joint compounds applied in coats that are too thick.

The procedure for repairing edge cracks depends on whether the crack is thin or wide. If the crack is small, coat it with a latex emulsion or a thin coat of joint or topping compound. Then sand as needed.

If the crack is wide, you may have to gouge out some of the joint compound to prepare the surface. Paint the gouged-out crack with a primer. Then fill with joint compound and sand smooth.

Center cracks— Cracks along the center of a joint occur for these reasons:

- If the joint tape is still intact, the crack is probably the result of applying joint compound too thick. Also, low humidity may have caused the compound to dry too quickly.

- If tape under the crack has been torn, it's possible that building settlement or movement caused the crack.

To repair cracks along the center of the joint, follow these steps:

1) If the joint tape is still intact and the crack is narrow, apply latex emulsion to the crack. If the crack is wide, use joint or topping compound to bridge the gap.

2) If the joint tape is torn, you may need to remove a section of tape and old joint compound before making repairs. Then retape the joint.

COMPOUND PROBLEMS

Compound has its own special set of problems. It can debond, grow mold, become pitted, sag and shrink. Let's look at each of these compound problems.

Debonding

When the joint or topping compound won't bond to (or becomes unbonded from) the joint tape or the drywall, this is known as *compound debonding*. Common causes of compound debonding include:

- A foreign substance (such as oil, dirt or sanding dust) on the drywall surface or on the surface of the joint tape when compound was applied.

- Improper mixing of the compound — the wrong ratio of water to dry mix.

- Using dirty water to mix the compound, or dirty tools to apply it.

- Using old compound.

Avoid most compound debonding by following the compound manufacturer's mixing instructions precisely. Some manufacturers request that you let the compound sit for a while after mixing it. There's a good reason for this. Don't take any shortcuts. They'll end up costing you time and money in the long run. Also, use only clean water to mix the compound, and clean tools to apply it. Be sure the drywall surface, the joint tape, and your mixing containers are clean as well.

Repairing compound debonding is very much like repairing tape blisters on a large scale. First, separate the debonded section of tape from the dried compound. Then remove enough of the old compound to allow you to apply a new layer in which to embed the tape. If the old compound crumbles easily, remove all of it. You'll also have to remove whatever compound was used to feather the joint. Apply new compound and tape as you would for a new joint.

Mold, Bacteria and Bad Odors

Contaminated water, dirty containers or tools, and letting the compound stand too long can result in mold, bacteria and bad odors in your compound. And remember that hot weather accelerates mold and bacteria growth.

If you discover your batch of compound has become a batch of unsavory brew, *throw it out.* Then soak your tools and containers in a chlorine bleach and water mixture overnight.

A professional drywall finisher is a neat and clean drywall finisher. Clean all your tools and equipment at the end of the day. Store compound to be used the next day in covered containers at room temperature.

Pitting

Small pits may appear in the finish of the compound after it dries. Pitting has three common causes:

- Air trapped in the compound mixture. This can happen if you mix the compound too hard.
- A compound mixture that's too thin.
- Not using enough pressure to apply the compound.

To prevent pitting, mix thoroughly with a slow, steady motion. If you're using a power mixer, run it at a slow speed. You're after a smooth mixture without any lumps. When applying the compound, use enough force both to bond it to the surface and to smooth it.

You can repair a section of pitted compound by simply filling in the pits. You may have to sand a bit to form a smooth base for applying the new compound. Then apply the new compound as you would apply a topping coat to the joint. Feather it out to conceal the joint. You may have to feather it wider than the original topping coat in order to fully hide the joint.

Sagging

Sagging compound or compound with runs in it occurs when:

- Compound is too thin. When mixed properly, compound is thick and smooth. Be sure to follow the manufacturer's mixing instructions precisely.

- Water added to the compound is too cold to mix completely.

To repair sags and runs, sand them smooth after they dry. Then recoat with joint or topping compound as needed.

Shrinking

If your compound shrinks excessively when it dries, you probably:

- Mixed the compound too thin.
- Didn't allow enough drying time between coats.
- Applied too much compound at one time.

You can remedy compound shrinkage by applying more compound. But be sure the previous coat is thoroughly dry before you begin any repair work.

DRYWALL PANEL PROBLEMS

Common problems in the panels themselves include drywall blisters, damaged edges, water damage, panel bowing, panel cracks, and fractures.

Drywall Blisters

When the facing paper comes unbonded from the surface of the drywall, it's known as a *drywall blister.* It may be caused by a manufacturing defect, or it may be the result of careless handling, which caused the gypsum filler to break inside the panel.

There are two common ways to repair drywall blisters:

1) Inject an aliphatic resin glue, such as white or yellow wood glue or carpenter's glue, into the blister and press the blister flat. This is the best remedy where the blister is small or where the blister isn't discovered until after the wall is textured and painted.

2) Cut out the entire blistered area and finish it with joint tape and joint compound. Follow the usual procedure for embedding joint tape and finishing the joint. If one width of joint tape isn't enough to cover the area, use as many strips as you need.

Damaged Edges

Improper handling of drywall panels may tear the facing paper and cause the gypsum core to crumble. To repair a damaged edge, cut off the damaged area back to sound drywall before you use the panel.

Water Damage

When a panel is wet, the core is soft and easily deformed. The facing paper may come unbonded from the panel. If a panel has been exposed to water, let it dry completely before installing it. Be sure it's thoroughly dry before you put it up.

Panel Bowing

Forcing a panel into a place that's too small will bow the panel. Remove the panel and trim the edges until it fits properly.

Panel Cracks and Fractures

A drywall panel can crack along its face or may fracture all the way through to the other side. Let's look at the causes of panel cracks and fractures and how to repair them.

Cracks— Cracks along the face of a drywall panel are most likely to show up over a doorway. This is a smaller and weaker section of panel. If a crack is over 1/8-inch wide, treat it just as you would a joint. Repair it by taping and feathering the joint compound or topping compound until the crack doesn't show.

This type of cracking is often caused by movement or settling of the building. Many larger buildings have built-in flexibility that may contribute to cracking. In a building with a flexible frame, the best choice is non-loadbearing gypsum partitions that have a clearance at the top of every wall. Fasten a runner (or track) to the ceiling to hold the tops of the metal studs. There can be up to 1/2-inch clearance between the wall and the ceiling. Fill this gap with a gasket or caulk. Place metal or plastic trim around the top edge of the drywall to give a finished appearance and add protection.

Fractures— There are three common reasons that drywall panels fracture:

- Improper handling.
- Attaching the panel across the wide face of structural members, such as headers. As the lumber shrinks, the panel is compressed and cracked.
- Scoring the facing paper beyond the edge of a cutout.

To repair a broken panel, cut out the damaged section and replace it. If the facing paper is scored beyond the edge of a cutout, repair it with joint tape.

Other Defects

This chapter covers the common defects that occur in a drywall installation. I've identified the probable causes and suggested repairs that work. But you may run into defects that aren't covered here. You can usually pinpoint the cause of the defect when you stop and think about the possibilities. The methods of repair are usually straightforward. When you decide on a repair, go after the root cause of the problem. Look beyond the surface of the defect. You don't want just a cosmetic fix: You want a fix that will stay fixed. You may have to remove part or all of the material — but it's worth it if you only have to repair it once.

The next chapter is on estimating. I'll describe step-by-step methods of estimating both material and labor for any size job. I'll also show you how to translate the cost of materials and labor into a bid that includes enough margin to cover your overhead, and leave you a profit at the end of the job.

Chapter 7

ESTIMATING DRYWALL WORK

There may be a few drywall contractors in the business because they enjoy the work, or like the creativity or the comradery or whatever. But most are in it for the money. Like any professional, drywall contractors like to be well paid for their time and trouble. If you hang and finish drywall for the fun of it, skip right on to the next chapter. There's nothing for you in the next few pages. But if you're in the drywall business to make a good living, this chapter should have your undivided attention.

The key to making money in the drywall contracting business is *knowing your costs and including every job cost in your estimate.* If you can predict the exact cost of materials, labor and overhead on a job, you can quote a competitive price that earns a good profit on the work you've done. That's your goal as a drywall estimator.

Drywall contractors who don't know their costs don't stay in business very long. If you can't estimate job costs with reasonable accuracy, one of two things will probably happen: Either you'll quote prices that are unreasonably high, and get little or no work, or quote prices that are too low, getting lots of work but losing your shirt. Either way, you won't survive long in the business.

Knowing your costs means knowing *all* your costs. That isn't just material and labor. It includes *all* the costs of running a business, including overhead, taxes and insurance. We'll cover these subjects in detail later in this chapter.

Once you know the cost of material, labor and overhead, you can add a percentage for contingency. No matter how careful and professional you may be as an estimator, there are going to be surprises — costs you couldn't foresee. Most surprises will increase your costs, not reduce them. A contingency allowance is intended to cover these costs that can't be predicted. For example, warped framing or a change in the weather may increase your labor cost. If it happens on most jobs, that makes it predictable and the cost should be in your estimate. If it happens only once in a while and can't be predicted, allow for it with contingency.

But note this about contingency. Including a big contingency allowance isn't a substitute for good estimating practice. Contingency should cover what you *can't estimate.* It isn't a slush fund used to cover mistakes and omissions. Do the very best you can to anticipate costs. Then add a few percent to cover the occasional cost that no one could predict.

The last item in most estimates will be *profit*. How much profit should you include in your estimates? The answer, of course, is as much as possible. But in practice, competition and good business judgment place limits on the profit you can earn on any one job.

Notice that I've already emphasized several times the importance of knowing your costs and including every job cost in your estimate. Here's why. *Every cost item left out of an estimate comes straight out of your profit*. You can underestimate labor costs by 5% or 10%, overestimate material costs or waste by several percent, misjudge overhead costs by a few percent. Those are mistakes any estimator can make — *and still make money on the job.* If you're lucky, the overestimates and underestimates will balance out. But forget to include the cost of drywall for the ceilings, or leave overhead or taxes out of your estimate, and your profit is gone! More likely, you'll do the job at a loss.

Here's my point. The only good estimate is a careful estimate that identifies every job cost and puts some number beside it. Some of your estimated costs will be a little too high. Others will be a little too low. That's not a serious problem. But your estimate on any cost item you forgot is always zero. That's a 100% miss. It doesn't take many mistakes like that to eat up your profit, your reason for doing the job.

Good estimators find ways to avoid costly errors. They've learned to include all job costs in their estimates. There's no point in running a business that loses money, even if you *can* pay your employees and keep the business limping along.

How do you remember to include every job cost in your estimate? Unfortunately, there's no single solution. Using a good checklist can help, of course. A checklist will remind you of cost items that might have been omitted. If you don't have a drywall estimating checklist, use mine. A copy is included later in this chapter.

Working systematically and being well-organized will eliminate many mistakes. Clear, well-organized estimates are usually good estimates. They make it easy for you, or anyone, to check the quantities and costs and verify how the estimate was compiled. They also make it easy to change, add or delete items as job requirements change. A good estimating form will help organize your estimates. If you don't have a good estimating form, use the one I use. A copy is included later in this chapter.

I recommend following some set procedure on every estimate. Good estimators are consistent estimators. They follow the same steps in the same order on every estimate. That makes omissions less likely. For most estimators, the order to follow in estimating is the same as the order the tradesmen will follow when doing the work. That helps the estimator visualize each step along the way.

Now, let's get down to brass tacks: setting up your own estimating system, a system that's both accurate and easy to use. First we'll look at eight key steps to follow when preparing an estimate. Then I'll explain how to develop the most accurate estimates possible for the types of work you handle. Next, we'll walk through the four main parts of an estimate: materials, labor, overhead and profit. We'll also look at payment terms, hidden costs and the proper procedure for submitting a bid. Finally, we'll compare our estimated costs to actual costs.

EIGHT IMPORTANT STEPS

There are eight key steps in making an estimate:

1) Make an appointment with the customer.

2) Review the job in detail with the customer. And review the plans and specifications in detail.

3) If the construction is far enough along for the framing to be complete, walk through to inspect the details and quality. If it's not possible to do a walk-through, find out as much as you can about the framing material and the contractor installing it. And find out whether you'll be required to repair framing defects. If so, be sure to allow for the necessary labor and materials in your estimate.

If possible, inspect framing done by the same crew in another location. This will at least give you an idea of the quality to expect on the job you're bidding. Of course, checking on the framer takes time. But it can save you time and money in the long run. Correcting stud and joist alignment can be time-consuming and *costly*. Reduce the unknowns in your estimate. Make an educated guess on the alignment problems you'll be facing.

4) Take good notes during your interview and your walk-through. No one has a perfect memory. And it's easy to forget important details that can cost you money later on. Use a good estimating form and your checklist to remind you of every cost in the job.

5) If possible, give the estimate to the customer at the time of the initial appointment. This will usually be possible on a small job such as a room addition or some patching work. Present the estimate at the end of the customer conference and you may get the job right on the spot!

If you can't submit the bid at the end of the interview, agree with your customer on a time when the bid should be submitted. I usually promise to deliver the estimate in 48 hours. That gives me enough time to make sure my bid is complete, accurate and profitable. If my customer needs a quote sooner, I can deliver sooner, of course. But two days is usually soon enough.

6) Submit your estimate on time. Many customers won't accept a late bid. And even if they do accept it, a late bid will make you look sloppy. This can cost you the job. Many builders and customers will assume that if you can't submit a bid on time, you probably can't stick to a construction schedule either.

7) When the customer accepts your bid, get his or her signature on the bid or on a contract form as soon as possible. This establishes a written commitment by the customer to have you do the work. It should also discourage unethical competitors from trying undercut your price.

8) Make sure the contract or bid covers key details. The starting date, if known, and the payment schedule are essential. *Don't neglect these details.* If you do, you'll find yourself with a completed job and only partial or no payment for your trouble.

CORNERSTONES OF A SOUND ESTIMATE

Preparing an accurate estimate takes time, even if you're an experienced professional estimator. Your first few estimates will take more time still. Eventually you'll learn shortcuts that save time without sacrificing accuracy. It takes practice and experience to perfect estimating skills. You may even become so skilled that you can make a reasonably accurate estimate based on the square footage of the building, the number of rooms, a quick walk-through and a review of the drawings and specifications. But never sacrifice accuracy for speed. A quick estimate is worthless unless it's accurate.

Keep Accurate Cost Records

The speed and accuracy of your estimates will increase even more if you review your *actual* costs after you complete each project. Compare your actual costs with your *estimated* costs. If actual costs don't match estimated cost, revise your material or manhour standards so the next estimate will be more accurate. If there was unnecessary waste of time or material on the job, find a way to improve productivity and reduce waste. Don't keep making the same mistake over and over again. Learn from your errors.

Use Detailed Estimating Forms

Let's look at some sample estimating forms. You'll probably use three estimating forms on every job: a material estimating form, a labor estimating form and a summary sheet.

Sample materials form— The first part of your estimate will be a survey of the materials needed. On most jobs, you determine material quantities by reading the plans and the job specifications. If the building is already framed, take your measurements from the building itself.

Your list of materials should include everything that's needed to complete the drywall installation. *Complete* means that the work is done as required by the specifications, drawings or any agreements reached with the customer. Your materials will include drywall panels, nails, screws or adhesives, joint tape, corner beading, expansion joints, joint compound, topping compound, and texture.

Figure 7-1 shows my combination material estimating form and checklist. I would use this form to estimate drywall in a new residence. Your estimating form doesn't have to be identical to this one. If this one doesn't quite suit your needs, develop a form that does. Just be sure that yours covers all the items you have to estimate. A good way to do this is to compare your checklist with the drawings and specs. Add to your checklist if necessary to cover all the work your jobs require.

Notice that my material takeoff form is designed for estimating just one room. My labor takeoff form is also designed for estimating just one room. I recommend that you use one material form and one labor form for each room, no matter how small the room.

Some drywall contractors have several checklists, one for each kind of job they handle: offices, apartment buildings, shopping malls, and so on. But once you find a form that works for you, stick with it. Consistency is essential for accurate estimating.

Sample labor form— The second part of your estimate for every room will be the labor cost. Labor includes *all* of the operations that must be performed by both the installation and

Room Material Estimate

Date:_____ **Sam's Sheetrock** **Page** ___ **of** ___

Job location/description__________________ **Room:**____________

Estimate for:_____________ **Room dimensions:**________________

Note: Material costs include 5 gallons of premixed joint compound (@ $_____/5 gal) and 380 linear feet of joint tape (@ $_____/500 foot roll) and 4½ lb of drywall nails (@ $_____/50 lb box) per 1,000 square feet of drywall. Includes 6% waste.

Item	Quantity	Size	Cost/Item	Cost
Ceiling:				
Std drywall	_____	½ x 4 x 8	$_____	= $_______
	_____	½ x 4 x 12	$_____	= $_______
WR drywall	_____	½ x 4 x 8	$_____	= $_______
	_____	½ x 4 x 12	$_____	= $_______
Type X	_____	⅝ x 4 x 8	$_____	= $_______
	_____	⅝ x 4 x 12	$_____	= $_______
Corner bead	_____		$_____/ft	= $_______
Expansion joint	_____		$_____/ft	= $_______
Texture @ 10 SF/lb =	___lbs		$_____/lb	= $_______
Other:	_____			= $_______
Ceiling material subtotal:				= $_______

Item	Quantity	Size	Cost/Item	Cost
Wall:				
Std drywall	_____	½ x 4 x 8	$_____	= $_______
	_____	½ x 4 x 12	$_____	= $_______
WR drywall	_____	½ x 4 x 8	$_____	= $_______
	_____	½ x 4 x 12	$_____	= $_______
Type X	_____	⅝ x 4 x 8	$_____	= $_______
	_____	⅝ x 4 x 12	$_____	= $_______
Corner bead	_____		$_____/ft	= $_______
Expansion joint	_____		$_____/ft	= $_______
Texture @ 10 SF/lb =	___lbs		$_____/lb	= $_______
Other:	_____			= $_______
Wall material subtotal:				= $_______
Room material subtotal:				= $_______

Materials checklist and estimate
Figure 7-1

finishing crews. Be sure to allow time for breaks and breakdowns, bad weather and the occasional crew that might not be quite as highly skilled or motivated as you thought they were. Figure 7-2 shows a detailed labor estimating form.

Sample summary form— If you're preparing an estimate for more than one room, you'll need detailed materials and labor forms for each room. Depending on the size of the job, the estimate may turn out to be many pages. To find the total job cost, you'll need to bring together material and labor costs from all sheets. That's the purpose of a summary sheet. It shows subtotals from each material and labor estimate sheet. If a room requires more than one page to estimate labor or materials, the last page should show the subtotal for that room. The room subtotal is then brought forward to the summary sheet. Figure 7-3 is an estimate summary sheet.

There are two advantages of using a summary sheet. First, it's easy for you to show a customer the cost for each room without going into too much detail. Second, if there's a major mistake in the material or labor estimate for one room, you're likely to see it when subtotals are compared on the summary sheet. Any subtotal that's much higher or lower than the rest may be wrong. Recheck your figures.

Notice that your estimate summary form also includes overhead, contingency and profit.

If you decide to design your own estimating forms, here are some key points to remember:

- Arrange the checklist in a logical sequence, usually the order you do the work. Recording your costs in a logical order makes it less likely that you'll leave anything out. You can see the job through the eyes of the installer, clearly visualizing each step in the process.

- Make your checklist (and all supporting calculations) neat and easy to read. It should be easy to go back and forth between the checklist and the calculations to check your figures or find a specific number.

- Make the checklist as detailed as possible. Check every item carefully. Don't leave anything to chance. You might not remember every important detail three weeks from now. So play it safe — list all important information right on your written estimate. Add margin notes if you think they'll be helpful later.

Room Labor Estimate

Date:_____ **Sam's Sheetrock** **Page____of____**

Job location/description:_________________ Room:_______________

Estimate for:______________ Room dimensions: ___________________

Labor

Ceiling: Install/tape _____ SF of panels x $_____/SF = $_______
Cut _____ openings @ .1 hrs/cut x $_____/hr = $_______
Install _____ ft of corner bead x $_____/ft = $_______
Install _____ ft of expansion joint x $_____/ft = $_______
Texture (SF x $_____/SF) = $_______
Repair of framing (LF x $_____/LF) = $_______
Install furring @ $_____/ft = $_______
Other: = $_______

Ceiling labor subtotal: = $_______

Walls: Install/tape _____ SF of panels x $_____/SF = $_______
Cut _____ openings @ .1 hrs/cut x $_____/hr = $_______
Install _____ ft of corner bead x $_____/ft = $_______
Install _____ ft of expansion joint x $_____/ft = $_______
Texture (SF x $_____/SF) = $_______
Repair of framing (LF x $_____/LF) = $_______
Install furring @ $_____/ft = $_______
Other: = $_______

Wall labor subtotal: = $_______

Room labor subtotal: = $_______

Labor estimating form
Figure 7-2

Estimate Summary

Date:____________

Page _____ **of** _____

Job location/description:______________________

Estimate for:________________________

Materials:

Room:	$__________
Room:	$__________
Room:	$__________
Room:	$__________
Room:	$__________
Room:	$__________
Room:	$__________
Room:	$__________
Room:	$__________

Materials **total:** **$**__________

Labor:

Room:	$__________
Room:	$__________
Room:	$__________
Room:	$__________
Room:	$__________
Room:	$__________
Room:	$__________
Room:	$__________
Room:	$__________

Labor **total:** **$**__________

Materials + labor:	$__________
Supervision:	$__________
Overhead:	$__________
Contingencies:	$__________
Total cost:	$__________
Profit:	$__________
Total estimate:	**$**__________

Summary checklist
Figure 7-3

ESTIMATING FROM BLUEPRINTS

There will be times when you'll have to prepare an estimate from plans, sometimes called *blueprints* or *prints* for short. Plans are construction drawings prepared by the building designer or architect. They show the design requirements for a construction project.

The plans have to give the builder enough information to complete the building. But most don't show every detail. All the information about material quality, sizes, and installation methods will be included in written specifications that are a supplement to the plans.

When reading plans, be sure to read all of the information on the drawings, including the explanatory notes. These notes should explain anything that's unclear on the drawings. Make sure you understand what you're bidding before completing the estimate. If you have questions, call the owner or designer. Don't bid a job if you don't understand what you're bidding.

Most plans are drawn on 36-inch by 24-inch sheets, with supplemental 8½-inch by 11-inch sheets. Large commercial projects normally have floor plans drawn to a scale of 1/8 inch equals 1 foot. Residential projects are usually drawn to a scale of 1/4 inch equals 1 foot. Different scales are used so that the plans will fit on and fill up most of a drawing sheet. The scale that's used will appear on the drawings.

Drawings often show dimensions as well as scale. Dimensions are normally shown as arrows between two points. The ends of the arrows may be defined by arrowheads, dots or diagonal slashes. Examples of dimension arrows are shown in Figure 7-4.

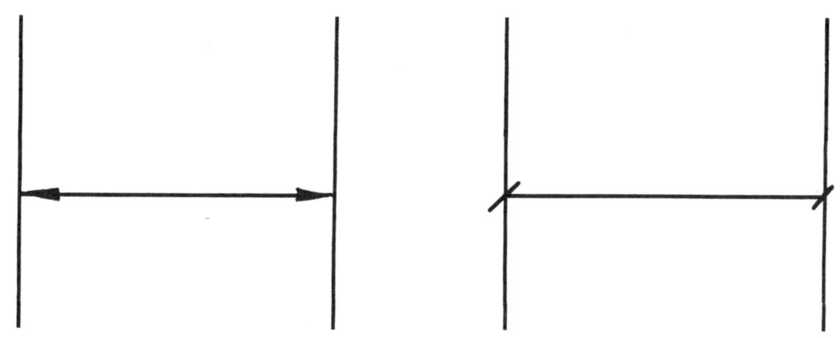

Examples of blueprint dimension arrows
Figure 7-4

Feet and inches are shown as numbers, but without the foot and inch marks. Where a dimension is shown as part of a note, foot and inch marks may be included. Elevation levels may be used to show where you are in the building.

Each sheet has a title block that shows the sheet number and what is covered on the drawing. For example, the title block might say "Sheet 2 of 6, Plan View, Second Floor." The title block is usually in the lower right-hand corner of the drawing. Each separate view on a sheet will be labeled. If there's a cross-section view of a double-hung window, it will be labeled as such. It may also say "typical" if it is representative of other double-hung window details.

When reading plans, be on the lookout for items marked "typical." This marks an item that appears many places in the drawings. The detail drawing of that item will be marked "typical" and shown only once. For example, if the same type of partition is used many places, the first time that partition occurs, it will be shown in a detail that's marked "typical." If there are other types of partitions, the first partition will probably be given a name, such as *Partition 1*. When that same partition occurs again in the drawings, it's simply marked Partition 1, with no detail shown. If there are other partitions in the drawings that are different from Partition 1, there will be another detail drawing or the differences will be noted.

The plans may also include a chart showing the finishes required for each room. See Figure 7-5.

Room	Acoustical ceiling	Stipple finish	Skip-texturing	Smooth wall
Living	x			x
Bed 1	x		x	
Bed 2	x		x	
Bed 3	x		x	
Den	x			x
Kitchen				x
Bath 1		x		
Bath 2		x		
Dining	x			x

Room finishes
Figure 7-5

Standard architectural abbreviations are often used in building drawings. Of course, the abbreviations one designer uses may not be the same as those another designer uses. Figure 7-6 shows some standard architectural abbreviations.

Blueprints for remodeling work are a little different from building plans for new construction. This is because remodeling work uses at least some part of the existing structure. Note carefully the parts of the existing structure that are to remain and the parts that are to be removed. Special symbols may be used to distinguish new and old work.

Remodeling plans may include a large number of notes detailing the work to be done. For this reason, the drawings are usually drawn to a scale of 1/4 inch equals 1 foot.

This isn't a book on plan reading, so I won't go into any more detail on how to understand what the plans show. If you need more help with plan reading, many books are available that cover the subject in detail. At the back of this manual you'll find an order form that lists several books that explain how to read plans.

ESTIMATING MATERIALS

I start most estimates by identifying each material item required. That information comes from the building plans and specs, of course. The specs or a note on the plans will identify the exact type and thickness of drywall for each surface to be covered. If you're installing specialty drywall that requires special joint tapes and compounds, be sure to include all of these items in your estimate.

Five essential drywall materials you'll show on your estimate are drywall panels, joint tape, corner bead, fasteners, and compound.

Drywall Panels

The standard 1/2-inch thick drywall panel is usually used for walls and ceilings where studs and ceiling joists are at 16 inches on center. Bathrooms and other high-moisture areas usually take 1/2-inch thick panels, but they need a water-resistant (WR) panel rather than the standard drywall panel. Always check the job specs before you decide on the type and thickness of panel.

Abbreviation	Definition
ACT	acoustical tile
ADJ	adjustable
BLDG	building
BLK	block
BR	bedroom
C-C	center-to-center
CMU	concrete masonry units
CONTR	contractor
DIM	dimension
DR	door
DWG	drawing
EL	elevation
ENC	enclose
EXT	exterior
F	face or Fahrenheit
FBD	fiberboard
FIN	finish
FUR	furred or furring
GA	gauge
GC	general contractor
GDW	gypsum drywall
HC	hollow core
HM	hollow metal
HT	height
INCL	included
ID	inside diameter
INT	interior
JST	joist
JT	joint
LAV	lavatory
LF	linear feet
LR	living room

Standard architectural abbreviations
Figure 7-6

Abbreviation	Definition
MFD	manufactured
MFR	manufacturer
MTL	metal
N	north
NO.	number
NTS	not to scale
OC	on center
OD	outside diameter
OPP	opposite
PBD	particle board
PERF	perforated
PLAS	plaster
PNL	panel
PSF	pounds per square foot
PSI	pounds per square inch
REF	reference
REV	revision
RM	room
S	south
SC	solid core
SCH	schedule
SF	square foot
SHT	sheet
STD	standard
T&G	tongue and groove
THK	thickness
TX	textured
TYP	typical
UNF	unfinished
VB	vapor barrier
VERT	vertical
W	west
WO	window opening

Standard architectural abbreviations (continued)
Figure 7-6

Room width	Room length 8'	10'	12'	14'	16'	18'	20'	22'
8'	320 SF	368 SF	416 SF	464 SF	512 SF	560 SF	608 SF	656 SF
9'	344 SF	394 SF	444 SF	494 SF	544 SF	594 SF	644 SF	694 SF
10'	368 SF	420 SF	472 SF	524 SF	576 SF	628 SF	680 SF	732 SF
11'	392 SF	446 SF	500 SF	554 SF	608 SF	662 SF	716 SF	770 SF
12'	416 SF	472 SF	528 SF	584 SF	640 SF	696 SF	752 SF	808 SF
13'	440 SF	498 SF	556 SF	614 SF	672 SF	730 SF	788 SF	846 SF
14'	464 SF	524 SF	584 SF	644 SF	704 SF	764 SF	824 SF	884 SF

Measuring surface area
Figure 7-7

Some jobs require 5/8-inch thick panels or some special type of panel.

When you're calculating the number of panels you need for the job, it's not enough just to figure the total square footage of the job and divide by 32. (There are 32 square feet in a 4 x 8 panel). You'll use a certain number of whole panels on each wall and ceiling surface. But you'll also be installing some partial panels. Here's where you can reduce waste of materials. Figure out how to install the panels so there's as little wasted material as possible — without doing excessive cutting, fitting and joint finishing.

For walls, you'll want to lay out the sheets with the long edges horizontal (perpendicular to the studs). This reduces the number of joints to be finished by about 25%. The exception to this layout is when the walls are more than 8 feet 1 inch high. In this case, use longer drywall panels and install them vertically (parallel to the studs).

To calculate the number of panels required for each room, follow these four steps:

1) Calculate the total surface area of each room, excluding the floor surface. Figure 7-7 shows surface area measurements for rooms of various lengths and widths. This table is based on

rooms that are rectangular and have ceilings 8 feet high. The dimensions in this table were calculated using the following formula:

Total room surface area (excluding floors)	=	length of room	x	width of room	+	2 (width x height)	+	2 (length x height)

2) Calculate the area of the doors and windows in the room. You'll deduct this figure from the total room surface. But don't bother calculating the area of windows smaller than 4 feet by 4 feet or doors smaller than 4 feet wide. The amount of material cut out for these openings probably can't be used elsewhere. Of course, there are occasionally cases where you can use small or odd-shaped pieces, such as around a small skylight opening.

For windows that are larger than 4 feet by 4 feet and doors that are wider than 3 feet, go ahead and calculate the area using one of the following formulas:

Area of rectangle	=	length x width
Area of square	=	length of any side^2
Area of circle	=	3.1416 x radius^2
Area of triangle	=	½ base x height

When you're measuring windows and doors, be sure to use the same units of measure that you used in your room surface area calculations. If you calculated the room surface area in feet, use feet to calculate the area of doors and windows. Don't use square yards for one and square feet for the other.

3) Now deduct the door and window areas from the total room surface area. This gives you the net surface area of the room. Here's the formula for net surface area.

Net surface area	=	total room surface area	—	area of doors and windows

4) Once you have the net surface area for a room, look at Figure 7-8. This figure shows the drywall coverage provided by

Number of panels	Panel size			
	4' x 8'	4' x 9'	4' x 10'	4' x 12'
10	320 SF	360 SF	400 SF	480 SF
11	352 SF	396 SF	440 SF	528 SF
12	384 SF	432 SF	480 SF	576 SF
13	416 SF	468 SF	520 SF	624 SF
14	448 SF	504 SF	560 SF	672 SF
15	480 SF	540 SF	600 SF	720 SF

Drywall coverage
Figure 7-8

various drywall panel sizes. You can expand this table to meet your needs. The table is based on the following formula:

$$\text{Panel coverage} = \text{number of panels} \times \text{panel length} \times \text{panel width}$$

Joint Tape and Corner Bead

Once you've calculated the number of drywall panels you need for a room, use the drywall layout to calculate the number of feet of panel joints. When you're figuring joint tape and corner bead requirements:

- Calculate the number of feet of flat joints and inside corner joints. You'll be using joint tape to finish these joints.

- Calculate the number of feet of outside corner joints. Outside corners require corner bead to make them strong and damage-resistant. If outside corners aren't on walls, but on such places as skylight openings, you may be able to use joint tape on the corner instead of bead. Before you make your final calculations, check the job specs to see what they require.

Drywall Fasteners and Joint Compound

Figure 7-9 shows the fastener spacing for screws and nails in laminated face ply and in screwed (or nailed) face-ply panels.

	Screw spacing Walls	Screw spacing Ceilings	Nail spacing Walls	Nail spacing Ceilings
Laminated face ply	16" o.c.	16" o.c.	8" o.c.	7" o.c.
Screwed or nailed face ply	24" o.c.	24" o.c.	16" o.c.	16" o.c.

Fastener spacing
Figure 7-9

Figure 7-10 shows the approximate quantities of fasteners and joint compound required per panel. This table is based on a 1/2-inch thick, 4 by 8 panel. For thicker or larger panels, you'll need to increase the quantity and length of nails or screws you use and the quantity of joint or texturing compound.

The values in Figure 7-10 are based on the following standards:

Nails for walls— Nineteen ring-shank nails, 1⅜ inch long, have a total weight of 1 ounce. The nail spacing is 8 inches. Stud spacing is 16 inches on center (o.c.). With horizontal panel installation, each panel is nailed to seven studs. There are seven nails per stud, or 49 nails per wall panel. The weight of the nails per 8-foot panel is 2.6 ounces (49 nails divided by 19 nails per ounce).

	Nails	Screws	Joint compound	Texture
Walls	2.6 oz/panel	1.75 oz/panel	48 lb/300 linear feet of joints	48 lb/200 square feet
Ceilings	3 oz/panel	2.2 oz/panel	48 lb/300 linear feet of joints	48 lb/200 square feet

Fastener and joint compound requirements
Figure 7-10

Nails for ceilings— Nineteen ring-shank nails, 1⅜-inch long, have a total weight of 1 ounce. The nail spacing is 7 inches. Joist spacing is 16 inches o.c. With perpendicular panel installation, each 8-foot panel is nailed to seven joists. There are eight nails per joist, or 56 nails per ceiling panel. The weight of the nails per 8-foot panel is 3 ounces (56 nails divided by 19 nails per ounce).

Screws for walls— Sixteen 1¼-inch long drywall screws weigh 1 ounce. Screw spacing is 16 inches. Stud spacing is 16 inches o.c. With horizontal panel installation, each 8-foot panel is screwed to seven studs. There are four screws per stud, or 28 screws per wall panel. The weight of the screws per 8-foot panel is 1.75 ounces (28 screws divided by 16 screws per ounce).

Screws for ceilings— Sixteen 1¼-inch long drywall screws weigh 1 ounce. Screw spacing is 12 inches. Joist spacing is 16 inches o.c. With perpendicular panel installation, each 8-foot panel is screwed to seven joists. There are five screws per joist, or 35 screws per ceiling panel. The weight of the screws per 8-foot panel is 2.2 ounces (35 screws divided by 16 screws per ounce).

Joint compound— Forty-eight pounds of ready-mix joint compound will cover about 300 linear feet (LF) of joints. Each 4 by 8 panel has 24 LF of edge. Edges of bottom panels on walls don't have to be taped and filled where the panel meets the floor. To adjust for these bottom panels, deduct 8 LF per panel. This leaves 16 LF of joints that actually have to be taped and filled for the bottom panels.

Here's an important point: When estimating, *do not double-count the perimeter joints.* If you count them with the ceiling, don't count them with the walls. Decide which surface you'll count them with, and stick to that with each estimate. That way you won't omit or double-count them.

Each pound of joint compound covers 6.25 LF (300 LF divided by 48 pounds). For all wall panels except bottom panels, each panel requires 3.84 pounds of joint compound (24 LF divided by 6.25 LF per pound). For bottom wall panels, each panel requires 2.56 pounds of compound (16 LF divided by 6.25 LF per pound).

Texture— Forty-eight pounds of texturing compound will cover 200 square feet (SF) at an average thickness of 1/32 inch. One 4 by 8 panel is 32 SF. The texturing for one panel requires 7.68 pounds of compound (48 pounds divided by 200 SF times 32 SF).

ESTIMATING LABOR

When you've completed your list of materials, you're ready to calculate the labor required to do the job. To calculate the labor cost, first figure out the number of labor hours required for the project. Then multiply the number of labor hours by the hourly labor rate to get the total labor cost. Here's what the formula looks like:

$$\text{Total labor cost} = \text{total number of labor hours} \times \text{labor rate per hour}$$

Your labor calculations will include all three of the major labor items: ceiling labor, wall labor, and finishing labor. But there's another important item to keep in mind when calculating the manhours: the quality of the framing. If the framing isn't plumb and level, you'll have to correct it before installing drywall. This takes time. So be sure to inspect the framing, if possible, and allow for any labor that might be required to align the framing.

Figure 7-11 is a table of sample manhour requirements for typical drywall installations. Of course your own labor figures, based on your crews and your job conditions, will be more accurate. Use my manhour figures only until you can compile good labor records of your own. They will also be useful as a check against gross errors in your estimates.

Ceiling Labor

Let's calculate the labor cost of installing six sheets of drywall on a ceiling. Assume, for this example, it takes two people 12 minutes to install one 4 x 8 ceiling panel, including fasteners but not including taping and finishing. The ceiling requires six of these panels, so we multiply 12 minutes by six panels to get 72

Work element	Manhours per 100 SF
Drywall on one face of metal or wood framing	
Nailed, 1 layer	
3/8"	1.8
1/2"	1.9
5/8"	2.1
Nailed, 2 layers	
3/8"	2.7
1/2"	3.0
5/8"	3.4
Screwed, 1 layer	
1/2"	1.6
5/8"	1.9
Glued, 1 layer	
1/2"	2.0
5/8"	1.9
Drywall for columns, pipe chases or fire partitions	
Nailed, 1 layer	
3/8"	4.4
1/2"	4.5
5/8"	4.6
Glued, 2 layers	
1/2"	8.5
5/8"	8.9
Glued, 3 layers	
1/2"	12.5
5/8"	13.0
Coreboard, 1 layer	
1-1/2"	4.0

Manhour tables
Figure 7-11

Work element	Manhours per 100 SF
Drywall for beams and soffits	
1 layer	
1/2"	4.0
5/8"	3.9
2 layers	
1/2"	7.3
5/8"	8.0
ADD additional time for:	
Ceiling work	.6
Walls over 9' high	.5
Resilient clip application	.4
Vinyl-covered drywall	.4
Veneer plaster finish	1.4
Textured spray	1.0
DEDUCT for no taping, finish or sanding	.9

Time includes moving on and off the site, unloading, stacking, installing drywall, taping, joint finishing, sanding, repair and clean-up as needed.

Manhour tables (continued)
Figure 7-11

minutes for the basic installation. To convert this to hours, divide by 60 minutes to get 1.2 labor hours. But two people are working on the basic installation, so we need to multiply the labor hours by 2 to get the total labor hours. This gives us a labor calculation of 2.4 hours for the basic ceiling installation. If the hourly labor rate were, say, \$20 an hour, then the labor cost would be \$48.

$$\text{Total labor hours} = \frac{\text{minutes per panel x number of panels}}{60} \times \text{number of workers}$$

$$= \frac{12 \times 6}{60} \times 2$$

$$= 1.2 \times 2$$

$$= 2.4 \text{ labor hours}$$

$$\text{Total labor cost} = \text{total labor hours} \times \text{hourly labor rate}$$
$$= 2.4 \times \$20$$
$$= \$48$$

Here's an alternate method of calculating labor costs. Divide the hourly labor rate ($20) by 60 minutes to get a per-minute labor cost of $0.33 per worker. Since there are two workers on this job, the per-minute cost is $0.33 times 2, or $0.66 per minute. Now just multiply the per-minute cost by the total number of minutes (72) required to do the work. This gives us a total labor cost of $47.52.

$$\text{Total labor cost} = \frac{\text{hourly labor rate}}{60} \times \text{number of workers} \times \text{total number of labor minutes}$$
$$= \frac{\$20}{60} \times 2 \times 72$$
$$= \$.33 \times 2 \times 72$$
$$= \$47.52$$

Notice that the first labor calculation method gave us a total labor cost of $48. You're probably wondering why the second method gave us a different answer than the first. Here's why. When we divided the hourly labor rate ($20) by 60 minutes, we got an answer of $0.33 per minute. This is because we rounded the answer off to two decimal places. If we'd used eight digits of accuracy instead of two, answer would have been $0.33333333. Since we lost a few cents by using the second method, it's better to use the first method in this case. Be aware of decimal points when you're calculating your costs.

Wall Labor

Once you've calculated the labor required for the ceiling, use the same procedure to calculate the labor required for the walls. Here are some variables to keep in mind when you're calculating wall labor.

If the framing is wood, you have the option of using either nails or screws for drywall installation, unless the customer has requested one or the other. Be sure to compare the material and labor costs of both screws and nails. If you use screws, fewer fasteners are required — so installing screws requires less labor. But screws cost more, so the material cost is higher. Screws have the advantage of far more holding power than nails. So, even with wood framing, there'll be less likelihood of fasteners coming loose, now or later, if you use screws instead of nails.

If you do use nails, the material cost will be lower but the labor cost will be higher. Be sure to consider these variables when installing wall panels on wood framing. If the framing is metal, you don't have to make any comparison between the cost of nails and the cost of screws. With metal framing, screws *must* be used.

In most walls, you'll have to cut one or more openings in the panels to allow for windows, doors, electrical outlets and switches. Make an accurate count of the cuts required and include the necessary labor time in your estimate.

The time required for installing corner bead must also be included in your estimate. Study the job specs to be sure nothing is left out of your labor estimate.

Finishing Labor

Finishing labor includes: taping and filling the joints, spotting the fastener heads, and applying the texturing or other finish material. Keep careful records of your finishing costs so you have accurate manhour figures for your estimates.

Square Foot and Linear Foot Costs

When you have accurate manhour figures for ceiling, wall and finishing labor, you can translate them into labor costs per square foot of drywall and per linear foot of corner bead. I've used square foot costs for the sample estimate later in the chapter.

LOOK OUT FOR HIDDEN COSTS

When preparing your estimate, be on the lookout for hidden costs. Even new construction projects have hidden costs you should be aware of. Let's look at four important items.

• Special fire resistance ratings: Using two thicknesses of 5/8-inch thick Type X drywall will usually provide the fire resistance rating that's required. But there are some ratings that require more thicknesses and special techniques. These situations will cost more money. So be sure to allow for them in your estimate.

• Specialized installations: For example, if you have strict sound isolation requirements, such as in a recording studio, your labor hours will increase. One of the best methods for isolating sound is to separate the framing of back-to-back wall surfaces. This means fastening the drywall to offset studs or to parallel double walls.

If you're only bidding on the drywall, the cost of installing offset studs or double wall framing isn't part of your estimate. But other requirements, such as close tolerance cuts and caulking, will increase your costs. Allow for these items in your estimate.

• Special texturing: Some textures are harder to put on than others, and so require more time. Skip-trowel (or skip-texture) is one of the fastest methods of applying texture to gypsum drywall. Most other types of texture take more time. For example, any texture that requires blotting will require more time than a trowel application. The more separate steps that are involved in a texture application, the more time that application will take. If you're bidding on a project that requires one of the more difficult textures, account for the additional labor in your bid.

• Down time: You won't always be able to go from one step directly to the next. Sometimes you'll have to pull off a job for several days and return later to finish the work. This increases your labor and transportation cost. For example, after the joint tape is embedded in the first layer of joint compound, you'll need to allow for drying time before applying the second coat. And you must allow additional time for the second coat of joint compound to dry before you can apply the third coat. And the third coat has to dry before you apply texture.

If the job is more than just one room, you can sequence operations from room to room to minimize down time. But if you're working on a small job, you'll have to stop work between coats. It may be possible to combine operations from different jobs to keep the crews working. But this means there will be time spent traveling between locations, and additional time needed for two setup and cleanup periods. In any case, costs will be higher than if work could go on without interruption. That's something to consider in your estimate.

Make sure the customer is informed in advance if there will be down time. Explain why down time is necessary. This is just good customer relations. If you don't take a few minutes to do it, the customer will assume, when you don't show up for a couple of days, that you're just doing his job when you have an odd moment from an important job. That can delay payment and almost certainly eliminate referrals.

YOUR "LABOR BURDEN"

When all labor and material quantities have been computed, it's time to begin figuring costs. For materials, that's easy. Just plug in the costs your dealer quotes for drywall and supplies. But for labor, converting manhours to costs is a little more complex. That's because a tradesman's hourly wage isn't the same as his hourly cost. The difference is the "labor burden." And that's a big difference. The "labor burden" will *add* between 25% and 30% to the labor cost on every drywall job. This means that for every dollar of payroll, you'll have to pay an additional 25 to 30 cents in taxes and insurance to government agencies and insurance companies.

Many drywall contractors routinely add 25% to their estimated labor cost to cover taxes and insurance. Here's a breakdown of where the "labor burden" comes from:

Unemployment insurance— All states levy an unemployment insurance tax on employers. This tax is based on total payroll for each calendar quarter and will vary with your history of unemployment claims. It may be as little as 1% or as much as 4% of payroll. You'll pay this tax monthly and file a state unemployment tax return quarterly.

The federal government also levies an unemployment insurance tax (FUTA). The tax has been about 0.8% of payroll up to a maximum per employee as established by law. This tax must be paid quarterly to a Federal Reserve Bank with Federal Tax Deposit Form 508. In January you file Form 940, showing all FUTA deposits made for the previous year, and pay any additional tax due.

Social Security and Medicare— The federal government also collects Social Security and Medicare (FICA) taxes. As an employer, your share is about 7½% of payroll on earnings up to an annual maximum per employee as set by law. Requirements vary, but you'll probably pay this tax several times a month. Deposits have to be made to a Federal Reserve Bank with Federal Tax Deposit Form 501. Every three months you'll file the Employer's Quarterly Federal Tax Return, Form 941, showing the FICA deposits made. These deposits must also include the *employee's* share of FICA tax and all income tax withheld from employee paychecks.

Workers' Comp— All states require that employers have Workers' Compensation Insurance to cover their employees in the event of a job-related injury. Heavy penalties are imposed on employers who fail to provide the required coverage. The cost of "workers' comp" insurance is taken as a percentage of payroll and is based on the type of work each employee performs. Clerical and office workers have a very low rate — they're not very likely to get hurt on the job. Your cost may be only a fraction of 1% of payroll. But hazardous occupations such as roofing carry a very high rate — usually at least 25% of payroll. The rate for most construction trades is between 5% and 10%. Drywall contractors usually have to pay about 8% to 10% of base pay. The actual cost will vary from year to year, depending on how many claims are filed. Your insurance carrier can quote the current cost of coverage.

Liability insurance— Every contractor should maintain liability insurance to protect the business from a lawsuit if there's an accident. On larger jobs the general contractor will probably require that you provide a certificate of insurance showing that you have the coverage required.

Liability insurance is also based on total payroll. The cost of coverage will vary by location, the liability limits needed, and your history of claims. But here's a rule of thumb that will be useful when estimating costs: Allow 5% of payroll for a comprehensive general liability policy, truck, auto and equipment floaters, fidelity bonds and umbrella liability coverage. Your insurance agent will quote the exact cost based on the coverage you need.

Here's a summary of the typical "labor burden"—

State Unemployment:	4.0%
Federal Unemployment (FUTA):	.8%
Social Security (FICA):	7.5%
Workers' Compensation:	10.0%
Liability Insurance:	5.0%

If you think that there's any way to cut corners on these taxes, you're mistaken. They have to be paid if you have employees and intend to stay in business. Both your state and the federal government have very effective ways to enforce payment of these taxes — and they apply heavy penalties against contractors who ignore the law.

That's why this labor burden has to be part of every estimate you make. If your labor burden is 25% and your wage cost for drywall will be $1,000 on a job, your labor cost will be $1,250 — $1,000 for wages and benefits and $250 for taxes and insurance.

OVERHEAD

Of course, wages, materials, taxes and insurance aren't your only costs on a job. Most of the rest is generally called *overhead*. It comes in two categories — direct and indirect.

Direct Overhead

You have some job costs that aren't labor, material or equipment. These costs are usually called *direct overhead*. These might include fire insurance, bid or performance bonds, permits, scaffolding, temporary water and electricity (if not furnished by the owner or general contractor) and repairs to adjoining property. You can probably think of many more direct overhead items. Some drywall contractors include in direct overhead the cost of supervision and other nonproductive labor such as the cost of estimating and selling the job. The time you spend on each job should be charged against that job. These are very real costs and must be included somewhere in the estimate. Since they're the result of taking each particular job, they can be properly included under direct overhead.

Indirect Overhead

After everything is figured, there are certain expenses you must bear in conducting your business, but which can't be charged directly against any single job. For example, office rent, telephone at your office, office lighting, office staff, small tools, office insurance, printing, your car, postage, advertising, and countless other items. These are indirect overhead expenses.

Some drywall contractors favor assigning the week's indirect overhead cost to the jobs that are in progress that week. For example, if you're doing two jobs of about the same size during a given week, each would bear one-half the indirect overhead cost for that week. Multiply weekly indirect overhead cost by the expected job duration to estimate indirect overhead cost for that project.

Some contractors load indirect overhead cost on each productive manhour. For example, if indirect overhead is $1,000 a month and your hangers and finishers work 1,000 manhours a month, the cost per manhour for indirect overhead would be $1. Other drywall contractors figure indirect overhead as a percentage of each bid. If contract volume averages $50,000 per month and indirect overhead is $2,500 per month, indirect overhead is 5%. I've even heard of contractors who have reduced indirect overhead expense to a cost per thousand square feet of drywall installed.

Any of these systems is good if it works. Keep a record of your indirect overhead and develop some method of dividing this cost among your jobs. Most important, don't forget to include this cost in your bid. Direct and indirect overhead together will be more than 10% of the estimated job cost for most drywall contractors. This 10% will make the difference between a profit and a loss on most of your jobs.

CONTINGENCY AND ESCALATION

As mentioned at the beginning of this chapter, most drywall contractors add a small amount to their bid to allow for problems that can't be forecast before work begins. The right amount to add for contingencies such as overtime, low productivity or misaligned framing depends on the job. For most drywall, 2% may be enough. Remodeling or repair work requires a larger allowance because it's harder to anticipate what's needed before work begins. Finally, if the plans are poorly drawn or unclear, the contingency allowance may have to be much higher.

Escalation covers increases in cost of labor, materials and equipment between the time the bid is submitted and the time work is actually done and paid for. Even if you're sure of the price of 1,000 square feet of drywall when making up the estimate, it may be hard to predict what panels will cost when actually ordered from the yard. Drywall prices can change rapidly. If you can't get firm quotes that are good for several weeks, it's good practice to either allow some amount for escalation, or exclude price increases from your bid.

PROFIT

Profit is your return on the investment in your contracting business. It isn't your wage. Profit is what's left after all expenses (including your wage) are paid. The business owner's income should be charged either to indirect overhead (when managing the business), to direct overhead (when supervising the work or estimating a job), or as a labor cost (when hanging board).

At the end of each year the profit earned should provide a reasonable return on money invested in the business (after you've drawn a reasonable wage). Profit should be 10% to 20% of the *tangible net worth* of your business. The tangible net worth is the value of everything the business owns, less what the business owes, and less any intangible items such as goodwill.

A drywall contractor with only a truck, some tools and a few hundred dollars in working capital may have a tangible net worth of about $10,000. Any profit left after taking a reasonable wage will very likely be used to buy more equipment and increase working capital. Still, that contractor should include enough profit in each bid to provide a 10% to 20% annual return on the $10,000 investment. That's $1,000 to $2,000 for the year.

Think of profit as interest on the money invested in equipment, office, inventory, work in progress, and everything else associated with running a drywall contracting business.

How much then, should you include in an estimate for profit? You'll hear contractors claim a 20% profit margin on some jobs. That's pretty high if it means that 20% is left after all costs (including the owner's wage) are covered. Some drywall contractors may operate efficiently enough to earn a 20% profit, but they're the exception. The contractor who boasts about a 20% profit probably means he has 20% left after paying for labor, materials and equipment. A good portion of the 20% that's left has to go toward overhead and the owner's wage. That isn't profit at all. A profit is what remains after *all* costs are considered, including overhead and the owner's wage.

What, then, is a realistic profit in the true sense? Dun and Bradstreet, the national credit reporting organization, has compiled figures on construction contractors for many years. They report the average profit after taxes for all contractors sampled to be consistently between 1.2% and 1.5% of gross receipts. This includes many contractors who reported losses or became insolvent. A 1.5% profit, even after taxes, is a fairly slim profit.

Not many drywall contractors, especially contractors on residential projects, include a 1% or 2% profit in their bids. On extremely large projects such as shopping centers or warehouses, a drywall contractor may include only 2% for profit — especially on a "cost plus" contract where risk of a loss is nearly eliminated.

Residential drywall contracting, especially remodeling and repair work, should carry a much higher profit margin because jobs are smaller and the risk of loss is larger. Probably 8% to 10% profit is a reasonable expectation on most jobs. Profit on small jobs or remodeling work may be as much as 25%. Of course, there's more to "profit" than just how much profit you would like to earn. Competition will usually limit the profit you can include in an estimate. If you include too much profit, you'll be underbid on work you would like to have. If you have enough work to stay busy and are asked to bid more, consider increasing the profit margin by a few percent.

When work is scarce, many drywall contractors take work at little or no profit just to keep their best crews busy (and themselves in the drywall contracting business). That's making the best of a bad situation. But it isn't a disaster if the bid you submit really does cover all your costs.

But before you cut down your profit, take a close look at your overhead costs. Your overhead may include items or services that you don't need at the present time and can do without. If you can reduce your overhead, you may be able to win a higher percentage of the available jobs without cutting your profit margin.

If, on the other hand, you're getting more work than you can handle, you can increase your profit margin slightly. But be sure to monitor your profit from job to job. And project a yearly average profit when you're doing your monitoring. You may need to adjust your profit percentage periodically to stay on target for your year-end profit goal.

To summarize, there is no single profit figure that fits all situations. For most drywall work, an 8% to 10% profit is reasonable. A contractor who has all the work he needs and wants, and is asked to bid on more, may feel that a 15% profit isn't excessive.

ESTIMATING REMODELING AND REPAIR WORK

Estimating remodeling and repair work is different from estimating new construction. For new construction, your estimate normally includes the installation and finishing of drywall. It may also include the installation of furring, and possibly, the squaring of framing. You base your quantities of material and the required labor hours solely on the surface area to be covered and the finish to be applied. That isn't the case when preparing an estimate for remodeling and repair work.

In remodeling and repair work, you'll have to match the new and existing wall or ceiling surfaces. This means you'll have to match the textures of the new and old surfaces. It also means that the two surfaces must be level with each other, even when one surface is drywall and the other is a plaster wall. It's always harder to match an existing wall or ceiling than it is to do a new installation. This is especially true if the section you're repairing is on the same surface as the existing work.

Slight differences in texture won't be particularly noticeable where the repaired or new surface is at right angles to the existing surface. But when the two surfaces are on the same plane, the match will have to be nearly exact. If it isn't matched carefully, the new work will be very noticeable. In fact, it'll look bad even if the new work is much better than the existing surface. Careful matching takes a lot of time and will increase your labor cost substantially. Be sure to allow for this.

To make sure the new and old surfaces are level with each other, first find out the thickness of the existing wall covering. If you have to match a plaster wall with drywall, there may not be a drywall thickness that will bring the two surfaces to the same level. You may have to apply veneer plaster to make the surface level.

If new framing has been added to a structure, you'll have to cover it so the new surface is level with the existing wall. If the new framing isn't exactly on the same plane as the existing framing, you may have to use furring and shimming to make the finished surface level.

Here's a helpful hint for matching new and old work: Sometimes it's easier to just retexture the old work right along with the new work. It may take more material, depending on the size of the job. But it may also save some labor and a great deal of frustration in the long run. Be sure to include the extra labor and materials in your estimate.

Remodeling and repair jobs can be good work. The profit margin is nearly always higher than new work. But it does require different skills and there are additional costs. Be sure to allow for them in your estimate.

CHECKING ESTIMATES

Your estimate isn't complete just because the bid price has been computed. Now the checking should begin. Every estimate you submit should be checked with the same care as the estimator used in preparing it. The best system is to have two estimators prepare estimates for the same job without comparing their figures until the work is done. You won't have enough time for that on most jobs. But, as a minimum, have someone check all price extensions and total all cost columns.

I've never met a drywall estimator who doesn't make mistakes — and you won't either. There are going to be errors in your estimates. Plan on it. Challenge whoever checks your work to find a significant mistake. Create an incentive to find errors before they become a major loss.

Checking is easier when you follow a consistent order, when your handwriting is clear and when you show all the computations used to arrive at totals. If you're making estimates on the back of an envelope in hieroglyphics that only a cryptographer could decipher, checking is nearly impossible.

Here are the most common estimating errors:

Errors in transcribing numbers— Every time you copy a number from one sheet to another, check to be sure copying was done correctly. Some estimators place a small check mark beside a number copied to another sheet, indicating that it has been transcribed and that it was copied correctly. Highlighter pens work well, too.

Mathematical errors— It's easy to add columns incorrectly or misplace a decimal point. Always check for errors in addition, multiplication, division and misplaced decimal points *before* submitting the bid.

Overlooked items— It's easy to overlook an item or two when preparing an estimate, especially on a large project. But even small omissions can be costly. Use a checklist or an estimating form that reduces your chance of making an error. Make a small check mark on the plans with colored pencil as each wall or ceiling area is estimated. Until all walls and ceilings have been checked, you've got more estimating to do. Use one estimating form for *materials* needed in each room and another for *labor* needed in each room. When complete, you should have an equal number of labor and material estimate sheets.

Inaccurate labor estimates— Some tradesmen are fast but do work that needs a lot of touching up. Others are slow but do perfect work. Most will fall somewhere in between the two extremes. To estimate labor requirements accurately, you have to know your tradesmen and their capabilities. Monitor job productivity carefully so you can predict future labor costs accurately.

Most major estimating errors will cost you either the job or some money. Either way, it's cash out of your pocket. There's no secret formula or magic trick to preparing error-free estimates. Just *prepare your bid carefully* and *base it on accurate cost records*. Use Figures 7-12 and 7-13 to help reduce errors in your estimates. Use these forms to record your estimated costs and your actual costs on each job after the job is complete. Refer back to these forms when you prepare your next estimate.

Date: ________

Job location/description: ________________________

Room: ________________

Room size: ____________

Amount ordered		Amount used	
___ ___ x ___ panels of _____		___ ___ x ___ panels of _____	
___ ___ x ___ panels of _____		___ ___ x ___ panels of _____	
Joint compound	_____ gallons		_____ gallons
Topping compound	_____ pounds		_____ pounds
Screws	_____ pounds		_____ pounds
Nails	_____ pounds		_____ pounds
Corner bead	_____ feet		_____ feet
Joint tape	_____ rolls		_____ rolls

Comparing estimated material with actual material
Figure 7-12

SUBMITTING YOUR BID

O.K. You've made detailed lists of materials and labor hours, taking into consideration every possible hidden cost. You've calculated your overhead cost and figured in a profit percentage. You've checked the bid for accuracy. Now it's time to get the bid accepted. Here are five important rules to remember when submitting your bid:

1) Get the bid in on time.

2) If the customer doesn't require that you use a specific bid form, use your own bid form. I use a modified version of my estimating form when bidding small jobs. If you bid on your own form, don't break out profit and overhead as separate items. Your profit and overhead are confidential. There's no need to share that information with competitors. Instead of listing profit and overhead separately, include a proportionate share of overhead and profit in every cost item.

Date:________

Job location/description:________________________

Room:________________

Room size:____________

Task	Estimated time	Actual time	Crew
Install wallboard	________hrs	________hrs	____________
Spot fastener holes and tape joints	________hrs	________hrs	____________
Finish joints	________hrs	________hrs	____________
Finish corners	________hrs	________hrs	____________
Texturing	________hrs	________hrs	____________
Other tasks	________hrs	________hrs	____________
Totals	________**hrs**	________**hrs**	

	Estimated time	Actual time	Crew
Room:	________hrs	________hrs	____________
Room:	________hrs	________hrs	____________
Room:	________hrs	________hrs	____________
Room:	________hrs	________hrs	____________
Job total	________**hrs**	________**hrs**	

	Estimated time	Actual time	Crew
Job:	________hrs	________hrs	____________
Job:	________hrs	________hrs	____________
Job:	________hrs	________hrs	____________
Job:	________hrs	________hrs	____________

Comparing estimated labor with actual labor
Figure 7-13

The smaller the job, the less formal your bid can be. If you're using the job contract as the bid form, make sure both you and your customer sign it.

3) Place a time limit on accepting the bid. For example, include a statement like: "This bid is void if not accepted within 14 days from this date." Otherwise your customer could accept the bid months later.

4) Get a written acceptance or written contract whenever possible. I violate this rule occasionally and haven't had it cost me any money yet. It probably will eventually. If a homeowner asks me to bid on installing drywall in his garage, I'll do the work without a written contract. I quote the price verbally, he accepts, we agree on a start date and terms of payment. That's a valid contract. But I'm taking a chance. If I have to sue in small claims court, it's my word against his.

Verbal contracts are binding to both parties. But they're dangerous. Settling disputes is harder when nothing is in writing. Use verbal contracts *only* when work will be done in a day or two and only if payment will be immediately upon completion. Even then, you're probably better off getting an agreement in writing. If there's any doubt about terms and payment, get your customer's signature on a contract like the sample, Figure 7-14.

5) When bidding larger jobs, be sure to get details on the bidding procedure before you begin work on the bid. On large jobs, the bidding process can be very complex. Find out what you're getting into before opening the plans. You don't want to lose a large contract because of some technicality, especially when you've put a lot of time into the estimate.

RECORDING AND CONTROLLING COSTS

Once your bid is accepted, the next step is to do the work — at the costs you estimated and with the profit you predicted. It's a nice pat on the back when estimated costs equal actual costs. But remember that it's possible to earn a second and third profit on every job. The first profit is the profit you estimated.

Proposal and Contract

For Residential Building Construction and Alteration

Date________________19____

To ____

Dear Sir:

We propose to furnish all material and perform all labor necessary to complete the following:

____

____

____

____

____

Job Location:

All of the above work to be completed in a substantial and workmanlike manner according to the drawings, job specifications, and terms and conditions on the back of this form for the sum of

Dollars ($________________________)

Payments to be made as the work progresses as follows:________________________________

____

____

the entire amount of the contract to be paid within_________days after substantial completion and acceptance by the owner. The price quoted is for immediate acceptance only. Delay in acceptance will require a verification of prevailing labor and material costs. This offer becomes a contract upon acceptance by contractor but shall be null and void if not executed within 5 days from the date above.

By________________________

"YOU, THE BUYER, MAY CANCEL THIS TRANSACTION AT ANY TIME PRIOR TO MIDNIGHT OF THE THIRD BUSINESS DAY AFTER THE DATE OF THIS TRANSACTION. SEE THE ATTACHED NOTICE OF CANCELLATION FORM FOR AN EXPLANATION OF THIS RIGHT."

You are hereby authorized to furnish all materials and labor required to complete the work according to the drawings, job specifications, and terms and conditions on the back of this proposal, for which we agree to pay the amounts itemized above

Owner ____

Owner ____ **Date**________________________

Accepted by Contractor______________________________ **Date**________________________

Sample installation contract
Figure 7-14

Notice To Customer Required By Federal Law

You have entered into a transaction on________________which may result in a lien, mortgage, or other security interest on your home. You have a legal right under federal law to cancel this transaction, if you desire to do so, without any penalty or obligation within three business days from the above date or any later date on which all material disclosures required under the Truth in Lending Act have been given to you. If you so cancel the transaction, any lien, mortgage, or other security interest on your home arising from this transaction is automatically void. You are also entitled to receive a refund of any down payment or other consideration if you cancel. If you decide to cancel this transaction, you may do so by notifying

____
(Name of Creditor)

at____
(Address of Creditor's Place of Business)

by mail or telegram sent not later than midnight of____________. You may also use any other form of written notice identifying the transaction if it is delivered to the above address not later than that time. This notice may be used for the purpose by dating and signing below.

I hereby cancel this transaction.

____________________ (Date) ____________________ (Customer's Signature)

Effect of rescission. When a customer exercises his right to rescind under paragraph (a) of this section, he is not liable for any finance or other charge, and any security interest becomes void upon such a rescission. Within 10 days after receipt of a notice of rescission, the creditor shall return to the customer any money or property given as earnest money, downpayment, or otherwise, and shall take any action necessary or appropriate to reflect the termination of any security interest created under the transaction. If the creditor has delivered any property to the customer, the customer may retain possession of it. Upon the performance of the creditor's obligations under this section, the customer shall tender the property to the creditor, except that if return of the property in kind would be impracticable or inequitable, the customer shall tender its reasonable value. Tender shall be made at the location of the property or at the residence of the customer, at the option of the customer. If the creditor does not take possession of the property within 10 days after tender by the customer, ownership of the property vests in the customer without obligation on his part to pay for it.

Sample Installation Contract (continued)
Figure 7-14

The second comes from reducing costs by careful purchasing and good supervision. The third profit is what you learn from comparing estimated and actual costs. Where are the estimating mistakes? What would you do differently on the next job? Don't miss this opportunity to record and control actual expenses.

All three profits require good cost-keeping — recording what's spent on each part of the job. A drywall contractor who has little or no payroll and personally watches every part of every job may need only a very simple cost-keeping system: A file for paid material bills on each job and a few notes on how many manhours were spent doing each task. But a larger drywall business needs a better system for recording and controlling costs.

The cost-keeping system you use should record what's spent on each part of every job: the number and cost of panels used of each type, the quantity and cost of fasteners, tape and joint compound, and the manhours needed to hang, tape and finish the board. Your records should be detailed enough to identify the labor and material cost for each square foot of drywall installed, for each linear foot of joint covered, and for each square foot of texture applied. These are unit cost items you'll use again and again when estimating.

Comparing estimated unit costs and actual unit costs will help you find estimating mistakes, of course. A comparison should also uncover waste of materials, poor supervision, less productive tradesmen, time lost waiting for supplies or tools, poor planning, excess labor and padded payrolls.

The job supervisor should consider your estimate to be the job budget. It should be a challenge to do the work with the manhours, materials and equipment listed in the estimate. If there are mistakes in the estimate, the supervisor isn't to blame, of course. But it's the supervisor's responsibility to bring any estimating errors to your attention as soon as they're noticed. The supervisor should try to deliver a finished job for the cost you estimated.

Your Cost File

Your estimates have to be based on your cost, your crews and the type of work you handle. No price taken from an estimating manual, no figure from another contractor, and nothing I can offer you is an adequate substitute for your cost records on jobs completed. No one can tell you what your costs are. You have to find that out yourself. That's why keeping good cost records is so important.

Collect copies of your completed estimates and summaries of actual job costs in a file or notebook. Probably the most convenient and practical book for this purpose is a three-ring 8½" x 11" binder. The advantage of a loose-leaf notebook is that it keeps all current estimates together. Yet you can remove an estimate at any time and take it with you without being burdened with the entire book. Also, any page can be removed and replaced if the plans change or when new work is added.

Each room should be estimated separately, as I suggested earlier, and the room estimates should be totaled on the summary sheet. This way, if you find an error in your estimate or changes are made in the plans, you only have to correct the section or sections affected and the summary sheet.

You may want to develop your own estimating forms or copy the forms included in this book. No matter what form you use, each estimate should be assigned a sequential number. Several sheets will be needed for each estimate, but all sheets of an estimate should bear the same estimate number. The first page of each estimate should show the estimate number, sheet number, address or lot number, owner, architect, date of bid, closing date, and the names of the estimator and the person who checked the estimate.

When your book is full, or as estimates are rejected, remove the unneeded estimates and file them in numerical order in a drawer or file box. An index in your file or in the estimate binder itself will simplify locating the cost data you're looking for.

This estimating file will become the most valuable and most used set of documents in your office. The time needed to hang, tape and texture each type of board on walls and ceilings in any size room should be recorded and filed for ready reference.

This cost data is essential when estimating, of course. But it also has other applications. Suppose your figures show that a two-man crew will usually take 40 hours to hang, tape and texture board in a typical home, including application of an acoustic ceiling finish. On the most recent job, your crew required 52 manhours. You would want to know what made this job more difficult or why the crew took so much more time.

TIPS ON ESTIMATING

There's a lot of information in this chapter. I've covered a lot of ground. But I can't end it without emphasizing eight key points that every estimator should understand:

1) Estimating requires precision and care. That's true today and always will be. There's no substitute for taking the time to work carefully.

2) When doing a takeoff, list materials in the order of construction. For example, hanging board should come before joint finishing.

3) Every estimate should be checked before it's accepted. Because checking is essential, every calculation should be easy to follow. The checker should be able to see at a glance how the estimator arrived at each cost listed. Where needed, include a sketch on the estimate showing the dimensions used.

4) When possible, use unit costs: find the cost per square foot and then multiply by the number of square feet. Compare unit costs with other estimates to verify that the figures are reasonable.

5) Check your figures. Don't use a figure that seems wrong.

6) Use only widely-accepted abbreviations. Abbreviations invented on the spot cause trouble.

7) Prefer decimals over fractions. Adding fractions causes mistakes. And the pocket calculators we all depend on can't handle fractions. Use a table of decimal equivalents to convert fractions to the decimal equivalent. Don't be concerned that there's no exact decimal equivalent for 2/3. The number 0.667 is close enough.

8) Be careful to distinguish between linear measure, square measure and cubic measure. For example, the tick marks used in writing linear feet (') and inches ('') should never be used for square feet or square inches. Twenty-five square feet should never be written *25 sq'*.

Care and attention to detail can make you a good drywall estimator. And behind every successful drywall contracting company is a good drywall estimator.

SAMPLE ESTIMATE

Let's do a step-by-step take-off of the materials and labor for installing drywall in a three-bedroom, two-bath residence. Figure 7-15 shows the plan view for the house we'll use. It shows dimensions for each of the rooms, as well as window and door sizes. All rooms have 8-foot ceilings.

To begin, we'll calculate the ceiling area for each room by multiplying the room length by the width. The master bedroom is 15 feet long and 12 feet wide. The ceiling area for the room is 180 square feet. If we use drywall panels that are 4 feet wide and 8 feet long for the ceiling, each panel will cover 32 square feet. To find out how many panels we'll need for the master bedroom ceiling, we'll divide 180 square feet by 32. The ceiling of this room will take 5.625 panels. Figures 7-16 and 7-17 show the two possible layouts for 4 x 8 panels. Since we can only buy whole panels, we'll have to buy six for this room.

But think again. Is that the best approach for this room? Note that the room is 12 feet wide. You can buy 12-foot long drywall panels that cover 48 square feet. If we divide 180 by 48, we find we only need 3.75 panels if we use the 12-foot length — so we could buy four panels instead of six. We'll have a strip left over that's 1 foot wide and 12 feet long. Figure 7-18 shows the arrangement. The 12-foot panels will take the same number of fasteners, but there will be fewer feet of joints. That's a better choice.

Now let's look at the master bedroom walls. One 15-foot wall has a 4 x 6 foot window. This wall can easily be covered using four 4 x 8-foot panels installed horizontally. The opposite 15-foot wall has so much surface area taken up by door openings that two 4 x 8 panels will nearly cover it. The approximately 4.5 square feet left over can be picked up from scrap pieces. Each of the 12-foot walls can be covered with two 12-foot panels. Figure 7-19 shows my material estimate for the master bedroom.

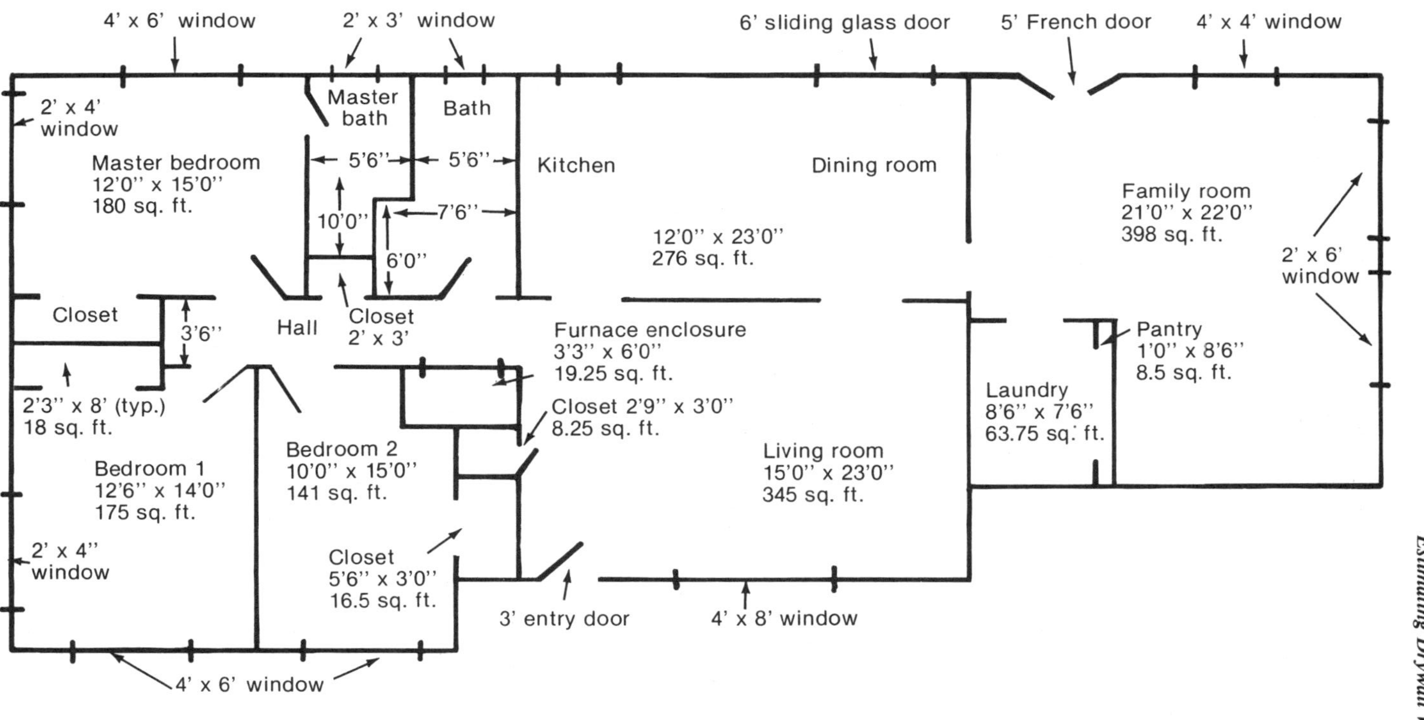

Plan view of sample residence
Figure 7-15

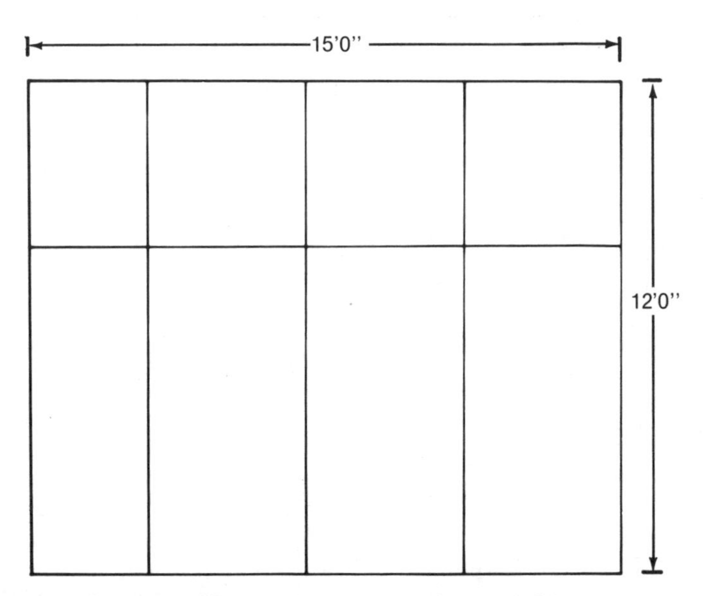

Master bedroom ceiling using 4' x 8' sheets laid in short dimension
Figure 7-16

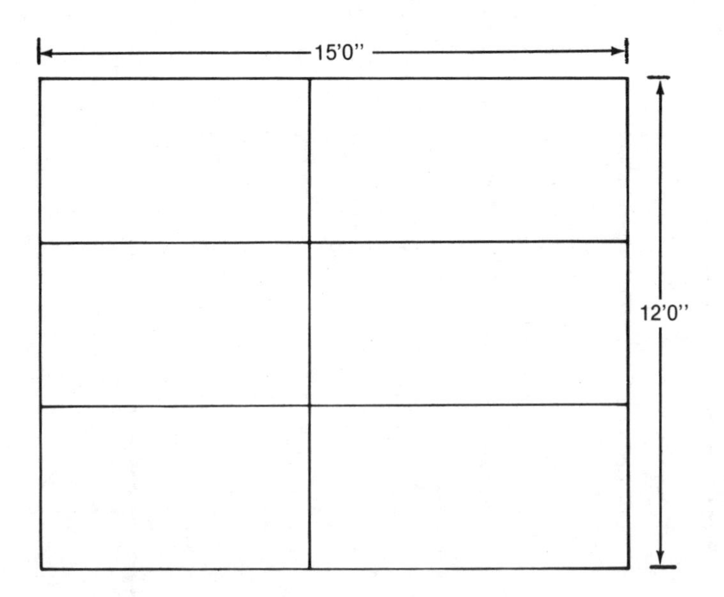

Master bedroom ceiling using 4' x 8' sheets laid on long dimension
Figure 7-17

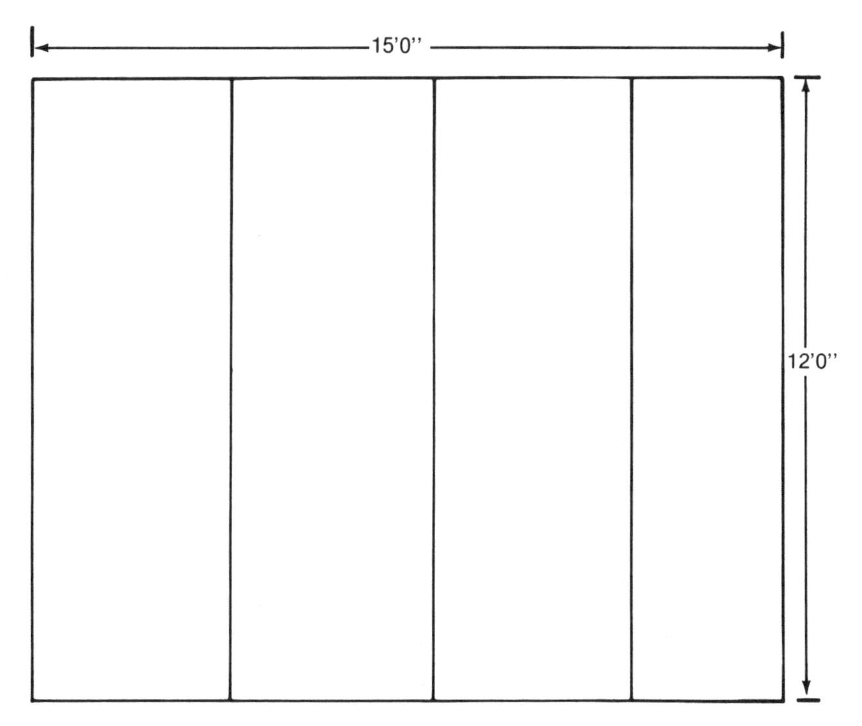

Master bedroom ceiling using 4' x 12' sheets
Figure 7-18

Now figure each of the other rooms and closets, using the same approach. Keep in mind the special materials, such as water resistant (WR) drywall for the bathrooms, and corner bead for outside corners. When you've completed the material and labor estimates, summarize the figures for each room on the material and labor summary sheet. Then add in a percentage for supervision, overhead, contingencies and profit.

Finally, you can check your estimate against mine. Figures 7-20 through 7-36 show my material and labor estimates for the nine rooms in our sample house. Figure 7-37 is my estimate summary.

Room Material Estimate

Date: 8/6/87 | Sam's Sheetrock | Page 1 of 9

Job location/description: RESIDENCE AT 4493 CEDAR LN | Room: MASTER BEDROOM AND CLOSET

Estimate for: J.B. SMITH | Room dimensions: 12'0" x 15'0" 2'3" x 8'0"

Note: Material costs include 5 gallons of premixed joint compound (@ $9.95/5 gal) and 380 linear feet of joint tape (@ $2.75/500 foot roll) and 4½ lb of drywall nails (@ $48.40/50 lb box) per 1,000 square feet of drywall. Includes 6% waste.

Item	Quantity	Size	Cost/Item	Cost
Ceiling:				
Std drywall	____	½ x 4 x 8	$____	= $____
	5	½ x 4 x 12	$8.93	= $44.65
WR drywall	____	½ x 4 x 8	$____	= $____
	____	½ x 4 x 12	$____	= $____
Type X	____	⅝ x 4 x 8	$____	= $____
	____	⅝ x 4 x 12	$____	= $____
Corner bead	____		$____/ft	= $____
Expansion joint	____		$____/ft	= $____
Texture @ 10 SF/lb =	20 lbs		$0.25/lb	= $5.00
Other:	____			= $____
Ceiling material subtotal:				= $49.65

Item	Quantity	Size	Cost/Item	Cost
Wall:				
Std drywall	10	½ x 4 x 8	$5.95	= $59.50
	4	½ x 4 x 12	$8.93	= $35.72
WR drywall	____	½ x 4 x 8	$____	= $____
	____	½ x 4 x 12	$____	= $____
Type X	____	⅝ x 4 x 8	$____	= $____
	____	⅝ x 4 x 12	$____	= $____
Corner bead	____		$____/ft	= $____
Expansion joint	____		$____/ft	= $____
Texture @ 10 SF/lb =	56 lbs		$0.25/lb	= $14.00
Other:	____			= $____
Wall material subtotal:				= $109.22
Room material subtotal:				= $158.87

Master bedroom material estimate
Figure 7-19

Room Material Estimate

Date: 8/6/87 Sam's Sheetrock Page 2 of 9

Job location/description: RESIDENCE AT 4493 CEDAR LN Room: BEDROOM 1 AND CLOSET

Estimate for: J.B. SMITH Room dimensions: 12'6" x 14'0" 2'3" x 8'8"

Note: Material costs include 5 gallons of premixed joint compound (@ $9.95/5 gal) and 380 linear feet of joint tape (@ $2.75/500 foot roll) and 4½ lb of drywall nails (@ $48.40/50 lb box) per 1,000 square feet of drywall. Includes 6% waste.

Item	Quantity	Size	Cost/Item	Cost
Ceiling:				
Std drywall	6	½ x 4 x 8	$5.95	= $35.70
		½ x 4 x 12	$	= $
WR drywall		½ x 4 x 8	$	= $
		½ x 4 x 12	$	= $
Type X		⅝ x 4 x 8	$	= $
		⅝ x 4 x 12	$	= $
Corner bead			$ /ft	= $
Expansion joint			$ /ft	= $
Texture @ 10 SF/lb =	19.5 lbs		$0.25/lb	= $4.88
Other:				= $
Ceiling material subtotal:				= $40.58

Item	Quantity	Size	Cost/Item	Cost
Wall:				
Std drywall	16	½ x 4 x 8	$5.95	= $95.20
		½ x 4 x 12	$	= $
WR drywall		½ x 4 x 8	$	= $
		½ x 4 x 12	$	= $
Type X		⅝ x 4 x 8	$	= $
		⅝ x 4 x 12	$	= $
Corner bead	8 FT		$0.17/ft	= $1.36
Expansion joint			$ /ft	= $
Texture @ 10 SF/lb =	51.2 lbs		$0.25/lb	= $12.80
Other:				= $
Wall material subtotal:				= $109.36
Room material subtotal:				= $149.94

Bedroom 1 material estimate
Figure 7-20

Room Material Estimate

Date: 8/6/87 Sam's Sheetrock Page 3 of 9

Job location/description RESIDENCE AT 4493 CEDAR LN Room: BEDROOM 2 AND CLOSET

Estimate for: J.B. SMITH Room dimensions: 10'0" x 15'0" 3'0" x 5'6"

Note: Material costs include 5 gallons of premixed joint compound (@ $9.95/5 gal) and 380 linear feet of joint tape (@ $2.75/500 foot roll) and 4½ lb of drywall nails (@ $48.40/50 lb box) per 1,000 square feet of drywall. Includes 6% waste.

Item	Quantity	Size	Cost/Item	Cost
Ceiling:				
Std drywall	5	½ x 4 x 8	$5.95	= $29.75
		½ x 4 x 12	$	= $
WR drywall		½ x 4 x 8	$	= $
		½ x 4 x 12	$	= $
Type X		⅝ x 4 x 8	$	= $
		⅝ x 4 x 12	$	= $
Corner bead			$ /ft	= $
Expansion joint			$ /ft	= $
Texture @ 10 SF/lb =	14.1 lbs		$0.25/lb	= $3.53
Other:				= $
Ceiling material subtotal:				= $33.28

Item	Quantity	Size	Cost/Item	Cost
Wall:				
Std drywall	16	½ x 4 x 8	$5.95	= $95.20
	2	½ x 4 x 12	$8.93	= $17.86
WR drywall		½ x 4 x 8	$	= $
		½ x 4 x 12	$	= $
Type X		⅝ x 4 x 8	$	= $
		⅝ x 4 x 12	$	= $
Corner bead	8 FT		$0.17/ft	= $1.36
Expansion joint			$ /ft	= $
Texture @ 10 SF/lb =	60.8 lbs		$0.25/lb	= $15.20
Other:				= $
Wall material subtotal:				= $129.62
Room material subtotal:				= $162.90

Bedroom 2 material estimate
Figure 7-21

Room Material Estimate

Date: 8/6/87 Sam's Sheetrock Page 4 of 9

Job location/description RESIDENCE AT 4493 CEDAR LN Room: 2 BATHROOMS

Estimate for: J.B. SMITH Room dimensions: 6'0" x 5'6" + 4'0" x 3'0" AND 6'0" x 5'6" + 6'0" x 7'6"

Note: Material costs include 5 gallons of premixed joint compound (@ $9.95/5 gal) and 380 linear feet of joint tape (@ $2.75/500 foot roll) and 4½ lb of drywall nails (@ $48.40/50 lb box) per 1,000 square feet of drywall. Includes 6% waste.

Item	Quantity	Size	Cost/Item	Cost
Ceiling:				
Std drywall	____	½ x 4 x 8	$____	= $____
	____	½ x 4 x 12	$____	= $____
WR drywall	1	½ x 4 x 8	$8.35	= $8.35
	3	½ x 4 x 12	$12.53	= $37.59
Type X	____	⅝ x 4 x 8	$____	= $____
	____	⅝ x 4 x 12	$____	= $____
Corner bead	____		$____/ft	= $____
Expansion joint	____		$____/ft	= $____
Texture @ 10 SF/lb =	12.3 lbs		$0.25/lb	= $3.08
Other:	____			= $____
Ceiling material subtotal:				= $49.02

Item	Quantity	Size	Cost/Item	Cost
Wall:				
Std drywall	____	½ x 4 x 8	$____	= $____
	____	½ x 4 x 12	$____	= $____
WR drywall	15	½ x 4 x 8	$8.35	= $125.25
	2	½ x 4 x 12	$12.53	= $25.06
Type X	____	⅝ x 4 x 8	$____	= $____
	____	⅝ x 4 x 12	$____	= $____
Corner bead	24 FT		$0.17/ft	= $4.08
Expansion joint	____		$____/ft	= $____
Texture @ 10 SF/lb =	28.2 lbs		$0.25/lb	= $7.05
Other:	____			= $____
Wall material subtotal:				= $161.44
Room material subtotal:				= $210.46

Bathrooms material estimate
Figure 7-22

Room Material Estimate

Date: 8/6/87 Sam's Sheetrock Page 5 of 9

Job location/description RESIDENCE AT 4493 CEDAR LN Room: HALL FURNACE CLOSET LINEN CLOSET

Estimate for: J. B. SMITH Room dimensions 18'6" x 3'6" + 2'0" x 3'0" + 3'3" x 6'0"

Note: Material costs include 5 gallons of premixed joint compound (@ $9.95/5 gal) and 380 linear feet of joint tape (@ $2.75/500 foot roll) and 4½ lb of drywall nails (@ $48.40/50 lb box) per 1,000 square feet of drywall. Includes 6% waste.

Item	Quantity	Size	Cost/Item	Cost
Ceiling:				
Std drywall	2	½ x 4 x 8	$5.95	= $11.90
	1	½ x 4 x 12	$8.93	= $8.93
WR drywall	____	½ x 4 x 8	$____	= $____
	____	½ x 4 x 12	$____	= $____
Type X	____	⅝ x 4 x 8	$____	= $____
	____	⅝ x 4 x 12	$____	= $____
Corner bead	____		$____/ft	= $____
Expansion joint	____		$____/ft	= $____
Texture @ 10 SF/lb =	9.03 lbs		$0.25/lb	= $2.26
Other:	____			= $____
Ceiling material subtotal:				= $23.09

Item	Quantity	Size	Cost/Item	Cost
Wall:				
Std drywall	14	½ x 4 x 8	$5.95	= $83.30
	____	½ x 4 x 12	$____	= $____
WR drywall	____	½ x 4 x 8	$____	= $____
	____	½ x 4 x 12	$____	= $____
Type X	____	⅝ x 4 x 8	$____	= $____
	____	⅝ x 4 x 12	$____	= $____
Corner bead	____		$____/ft	= $____
Expansion joint	____		$____/ft	= $____
Texture @ 10 SF/lb =	44 lbs		$0.25/lb	= $11.00
Other:	____			= $____
Wall material subtotal:				= $94.30
Room material subtotal:				= $117.39

Hall material estimate
Figure 7-23

Room Material Estimate

Date: 8/6/87 Sam's Sheetrock Page 6 of 9

Job location/description RESIDENCE AT 4493 CEDAR LN Room: LIVING ROOM AND CLOSET

Estimate for: J.B. SMITH Room dimensions: 15'0" x 23'0" + 3'0" x 5'6"

Note: Material costs include 5 gallons of premixed joint compound (@ $9.95/5 gal) and 380 linear feet of joint tape (@ $2.75/500 foot roll) and 4½ lb of drywall nails (@ $48.40/50 lb box) per 1,000 square feet of drywall. Includes 6% waste.

Item	Quantity	Size	Cost/Item	Cost
Ceiling:				
Std drywall	1	½ x 4 x 8	$5.95	= $5.95
	8	½ x 4 x 12	$8.93	= $71.44
WR drywall	____	½ x 4 x 8	$____	= $____
	____	½ x 4 x 12	$____	= $____
Type X	____	⅝ x 4 x 8	$____	= $____
	____	⅝ x 4 x 12	$____	= $____
Corner bead	____		$____/ft	= $____
Expansion joint	____		$____/ft	= $____
Texture @ 10 SF/lb =	35.4 lbs		$0.25/lb	= $8.85
Other:	____			= $____
Ceiling material subtotal:				= $86.24

Item	Quantity	Size	Cost/Item	Cost
Wall:				
Std drywall	14	½ x 4 x 8	$5.95	= $83.30
	4	½ x 4 x 12	$8.93	= $35.72
WR drywall	____	½ x 4 x 8	$____	= $____
	____	½ x 4 x 12	$____	= $____
Type X	____	⅝ x 4 x 8	$____	= $____
	____	⅝ x 4 x 12	$____	= $____
Corner bead	8 FT		$0.17/ft	= $1.36
Expansion joint	____		$____/ft	= $____
Texture @ 10 SF/lb =	32.5 lbs		$0.25/lb	= $8.13
Other:	____			= $____
Wall material subtotal:				= $128.51
Room material subtotal:				= $214.75

Living room material estimate
Figure 7-24

Room Material Estimate

Date: 8/6/87 Sam's Sheetrock Page 7 of 9

Job location/description: RESIDENCE AT 4493 CEDAR LN Room: KITCHEN/DINING ROOM

Estimate for: J.B. SMITH Room dimensions: 12'0" x 23'0"

Note: Material costs include 5 gallons of premixed joint compound (@ $9.95/5 gal) and 380 linear feet of joint tape (@ $2.75/500 foot roll) and 4½ lb of drywall nails (@ $48.40/50 lb box) per 1,000 square feet of drywall. Includes 6% waste.

Item	Quantity	Size	Cost/Item	Cost
Ceiling:				
Std drywall	____	½ x 4 x 8	$____	= $____
	6	½ x 4 x 12	$8.93	= $53.58
WR drywall	____	½ x 4 x 8	$____	= $____
	____	½ x 4 x 12	$____	= $____
Type X	____	⅝ x 4 x 8	$____	= $____
	____	⅝ x 4 x 12	$____	= $____
Corner bead	____		$____/ft	= $____
Expansion joint	____		$____/ft	= $____
Texture @ 10 SF/lb =	27.6 lbs		$0.25/lb	= $6.90
Other:	____			= $____
Ceiling material subtotal:				= $60.48

Item	Quantity	Size	Cost/Item	Cost
Wall:				
Std drywall	____	½ x 4 x 8	$____	= $____
	10	½ x 4 x 12	$8.93	= $89.30
WR drywall	____	½ x 4 x 8	$____	= $____
	2	½ x 4 x 12	$12.53	= $25.06
Type X	____	⅝ x 4 x 8	$____	= $____
	____	⅝ x 4 x 12	$____	= $____
Corner bead	____		$____/ft	= $____
Expansion joint	____		$____/ft	= $____
Texture @ 10 SF/lb =	44.3 lbs		$0.25/lb	= $11.08
Other:	____			= $____
Wall material subtotal:				= $125.44
Room material subtotal:				= $185.92

Kitchen/dining room material estimate
Figure 7-25

Room Material Estimate

Date: 8/6/87 Sam's Sheetrock Page 8 of 9

Job location/description RESIDENCE AT 4493 CEDAR LN Room: FAMILY ROOM

Estimate for: J.B. SMITH Room dimensions: 21'0" x 22'0" + 8'6" x 8'6"

Note: Material costs include 5 gallons of premixed joint compound (@ $9.95/5 gal) and 380 linear feet of joint tape (@ $2.75/500 foot roll) and 4½ lb of drywall nails (@ $48.40/50 lb box) per 1,000 square feet of drywall. Includes 6% waste.

Item	Quantity	Size	Cost/Item	Cost
Ceiling:				
Std drywall	2	½ x 4 x 8	$5.95	= $11.90
	3	½ x 4 x 12	$8.93	= $26.79
WR drywall	____	½ x 4 x 8	$____	= $____
	____	½ x 4 x 12	$____	= $____
Type X	____	⅝ x 4 x 8	$____	= $____
	____	⅝ x 4 x 12	$____	= $____
Corner bead	____		$____/ft	= $____
Expansion joint	____		$____/ft	= $____
Texture @ 10 SF/lb =	38.8 lbs		$0.25/lb	= $9.70
Other:	____			= $____
Ceiling material subtotal:				= $48.39

Item	Quantity	Size	Cost/Item	Cost
Wall:				
Std drywall	13	½ x 4 x 8	$5.95	= $77.35
	8	½ x 4 x 12	$8.93	= $71.44
WR drywall	____	½ x 4 x 8	$____	= $____
	____	½ x 4 x 12	$____	= $____
Type X	____	⅝ x 4 x 8	$____	= $____
	____	⅝ x 4 x 12	$____	= $____
Corner bead	8 FT		$0.17/ft	= $1.36
Expansion joint	____		$____/ft	= $____
Texture @ 10 SF/lb =	77.7 lbs		$0.25/lb	= $19.43
Other:	____			= $____
Wall material subtotal:				= $169.58
Room material subtotal:				= $217.97

Family room material estimate
Figure 7-26

Room Material Estimate

Date: 8/6/87 Sam's Sheetrock Page 9 of 9

Job location/description RESIDENCE AT 4493 CEDAR LN Room: LAUNDRY & PANTRY

Estimate for: J.B. SMITH Room dimensions: 8'6" x 7'6" AND 8'6" x 1'0"

Note: Material costs include 5 gallons of premixed joint compound (@ $9.95/5 gal) and 380 linear feet of joint tape (@ $2.75/500 foot roll) and 4½ lb of drywall nails (@ $48.40/50 lb box) per 1,000 square feet of drywall. Includes 6% waste.

Item	Quantity	Size	Cost/Item	Cost
Ceiling:				
Std drywall	____	½ x 4 x 8	$____	= $____
	2	½ x 4 x 12	$8.93	= $17.86
WR drywall	____	½ x 4 x 8	$____	= $____
	____	½ x 4 x 12	$____	= $____
Type X	____	⅝ x 4 x 8	$____	= $____
	____	⅝ x 4 x 12	$____	= $____
Corner bead	____		$____/ft	= $____
Expansion joint	____		$____/ft	= $____
Texture @ 10 SF/lb =	7.3 lbs		$0.25/lb	= $1.83
Other:	____			= $____
Ceiling material subtotal:				= $19.69

Item	Quantity	Size	Cost/Item	Cost
Wall:				
Std drywall	6	½ x 4 x 8	$5.95	= $35.70
	____	½ x 4 x 12	$____	= $____
WR drywall	3	½ x 4 x 8	$8.35	= $25.05
	____	½ x 4 x 12	$____	= $____
Type X	____	⅝ x 4 x 8	$____	= $____
	____	⅝ x 4 x 12	$____	= $____
Corner bead	____		$____/ft	= $____
Expansion joint	____		$____/ft	= $____
Texture @ 10 SF/lb =	28.8 lbs		$0.25/lb	= $7.20
Other:	____			= $____
Wall material subtotal:				= $67.95
Room material subtotal:				= $87.64

Laundry and pantry material estimate
Figure 7-27

Room Labor Estimate

Date: 8/6/87 Sam's Sheetrock Page 1 of 9

Job location/description: RESIDENCE AT 4493 CEDAR LN Room: MASTER BEDROOM AND CLOSET

Estimate for: J.B.SMITH Room dimensions: 12'0" X 15'0"

Labor

Ceiling:	Install/tape 198 SF of panels x $.462/SF	= $91.48
	Cut _____ openings @ .1 hrs/cut x $_____/hr	= $_____
	Install _____ ft of corner bead x $_____/ft	= $_____
	Install _____ ft of expansion joint x $_____/ft	= $_____
	Texture (SF x $.15 /SF)	= $ 29.70
	Repair of framing (LF x $_____/LF)	= $_____
	Install furring @ $_____/ft	= $_____
	Other:	= $_____
Ceiling labor subtotal:		= $121.18
Walls:	Install/tape 512 SF of panels x $.351/SF	= $179.71
	Cut 6 openings @ .1 hrs/cut x $20 /hr	= $ 12.00
	Install 8 ft of corner bead x $.56/ft	= $ 4.48
	Install _____ ft of expansion joint x $_____/ft	= $_____
	Texture (SF x $.12 /SF)	= $ 61.44
	Repair of framing (LF x $_____/LF)	= $_____
	Install furring @ $_____/ft	= $_____
	Other:	= $_____
Wall labor subtotal:		= $257.63
Room labor subtotal:		= $378.81

Master bedroom labor estimate
Figure 7-28

Room Labor Estimate

Date: 8/6/87   Sam's Sheetrock   Page 2 of 9

Job location/description: RESIDENCE AT 4493 CEDAR LN   Room: BEDROOM 1 AND CLOSET

Estimate for: J. B. SMITH   Room dimensions: 12'6" x 14'0" + 2'3" x 8'0"

Labor

Ceiling:	Install/tape 192 SF of panels x $.462/SF	= $88.70
	Cut ____ openings @ .1 hrs/cut x $____/hr	= $____
	Install ____ ft of corner bead x $____/ft	= $____
	Install ____ ft of expansion joint x $____/ft	= $____
	Texture (SF x $.15/SF)	= $28.80
	Repair of framing (LF x $____/LF)	= $____
	Install furring @ $____/ft	= $____
	Other:	= $____
Ceiling labor subtotal:		= $117.50
Walls:	Install/tape 512 SF of panels x $.351/SF	= $179.71
	Cut 7 openings @ .1 hrs/cut x $20/hr	= $14.00
	Install 8 ft of corner bead x $.56/ft	= $4.48
	Install ____ ft of expansion joint x $____/ft	= $____
	Texture (SF x $.12/SF)	= $61.44
	Repair of framing (LF x $____/LF)	= $____
	Install furring @ $____/ft	= $____
	Other:	= $____
Wall labor subtotal:		= $259.63
Room labor subtotal:		= $377.13

Bedroom 1 labor estimate
Figure 7-29

Room Labor Estimate

Date: 8/6/87 Sam's Sheetrock Page 3 of 9

Job location/description: RESIDENCE AT 4493 CEDAR LN Room: BEDROOM 2 AND CLOSET

Estimate for: J.B. SMITH Room dimensions: 10'0" x 15'0" + 3'0" x 5'6"

Labor

Ceiling:	Install/tape 141 SF of panels x $.462/SF	= $65.14
	Cut 1 openings @ .1 hrs/cut x $20/hr	= $2.00
	Install 8 ft of corner bead x $.56/ft	= $4.48
	Install ___ ft of expansion joint x $___/ft	= $___
	Texture (SF x $.15/SF)	= $21.15
	Repair of framing (LF x $___/LF)	= $___
	Install furring @ $___/ft	= $___
	Other:	= $___
Ceiling labor subtotal:		= $92.77
Walls:	Install/tape 608 SF of panels x $.351/SF	= $213.41
	Cut 7 openings @ .1 hrs/cut x $20/hr	= $14.00
	Install 8 ft of corner bead x $.56/ft	= $4.48
	Install ___ ft of expansion joint x $___/ft	= $___
	Texture (SF x $.12/SF)	= $72.96
	Repair of framing (LF x $___/LF)	= $___
	Install furring @ $___/ft	= $___
	Other:	= $___
Wall labor subtotal:		= $304.85
Room labor subtotal:		= $397.62

Bedroom 2 labor estimate
Figure 7-30

Room Labor Estimate

Date: 8/6/87 | Sam's Sheetrock | Page 4 of 9

Job location/description: RESIDENCE AT 4493 CEDAR LN Room: 2 BATHROOMS

Estimate for: J.B. SMITH Room dimensions: 6'0" x 5'6" + 4'0" x 3'0" 6'0" x 5'6" + 6'0" x 7'6"

Labor

Ceiling:
Install/tape 123 SF of panels x $.462/SF = $56.83
Cut 2 openings @ .1 hrs/cut x $20/hr = $4.00
Install ___ ft of corner bead x $___/ft = $___
Install ___ ft of expansion joint x $___/ft = $___
Texture (SF x $.15/SF) = $18.45
Repair of framing (LF x $___/LF) = $___
Install furring @ $___/ft = $___
Other: = $___

Ceiling labor subtotal: = $79.28

Walls:
Install/tape 282 SF of panels x $.351/SF = $98.98
Cut 15 openings @ .1 hrs/cut x $20/hr = $30.00
Install 24 ft of corner bead x $.56/ft = $13.44
Install ___ ft of expansion joint x $___/ft = $___
Texture (SF x $.12/SF) = $33.84
Repair of framing (LF x $___/LF) = $___
Install furring @ $___/ft = $___
Other: = $___

Wall labor subtotal: = $176.26

Room labor subtotal: = $255.54

Bathrooms labor estimate
Figure 7-31

Room Labor Estimate

Date: 8/6/87 Sam's Sheetrock Page 5 of 9

Job location/description: RESIDENCE AT 4493 CEDAR LN Room: HALL, FURNACE CLOSET, LINEN CLOSET

Estimate for: J.B. SMITH Room dimensions: 18'6" x 3'6" 2'0" x 3'0" 3'3" x 6'0"

Labor

Ceiling: Install/tape 90.25 SF of panels x $.462/SF = $ 41.70
Cut 1 openings @ .1 hrs/cut x $ 20 /hr = $ 2.00
Install ____ ft of corner bead x $____/ft = $____
Install ____ ft of expansion joint x $____/ft = $____
Texture (SF x $.15 /SF) = $ 13.54
Repair of framing (LF x $____/LF) = $____
Install furring @ $____/ft = $____
Other: = $____

Ceiling labor subtotal: = $ 57.24

Walls: Install/tape 440 SF of panels x $.351 /SF = $ 154.44
Cut 8 openings @ .1 hrs/cut x $ 20 /hr = $ 16.00
Install ____ ft of corner bead x $____/ft = $____
Install ____ ft of expansion joint x $____/ft = $____
Texture (SF x $.12 /SF) = $ 52.80
Repair of framing (LF x $____/LF) = $____
Install furring @ $____/ft = $____
Other: = $____

Wall labor subtotal: = $ 223.24

Room labor subtotal: = $ 280.48

Hall labor estimate
Figure 7-32

Room Labor Estimate

Date: 8/6/87 Sam's Sheetrock Page 6 of 9

Job location/description: RESIDENCE AT 4493 CEDAR LN Room: LIVING ROOM AND CLOSET

Estimate for: J.B. SMITH Room dimensions: 15'0" X 23'0" + 3'0" X 2'9"

Labor

Ceiling:	Install/tape 354 SF of panels x $.462/SF	= $163.55
	Cut ____ openings @ .1 hrs/cut x $____/hr	= $____
	Install ____ ft of corner bead x $____/ft	= $____
	Install ____ ft of expansion joint x $____/ft	= $____
	Texture (SF x $.15 /SF)	= $53.10
	Repair of framing (LF x $____/LF)	= $____
	Install furring @ $____/ft	= $____
	Other:	= $____
Ceiling labor subtotal:		= $216.65
Walls:	Install/tape 325 SF of panels x $.351/SF	= $114.00
	Cut 8 openings @ .1 hrs/cut x $20/hr	= $16.00
	Install 8 ft of corner bead x $.56/ft	= $4.48
	Install ____ ft of expansion joint x $____/ft	= $____
	Texture (SF x $.12 /SF)	= $39.00
	Repair of framing (LF x $____/LF)	= $____
	Install furring @ $____/ft	= $____
	Other:	= $____
Wall labor subtotal:		= $173.48
Room labor subtotal:		= $390.13

Living room labor estimate
Figure 7-33

Room Labor Estimate

Date: 8/6/87 Sam's Sheetrock Page 7 of 9

Job location/description: RESIDENCE AT 4493 CEDAR LN Room: KITCHEN/ DINING ROOM

Estimate for: J.B. SMITH Room dimensions: 12'0" X 23'0"

Labor

Ceiling:	Install/tape 276 SF of panels x $.462/SF	= $127.51
	Cut 2 openings @ .1 hrs/cut x $20 /hr	= $4.00
	Install ____ ft of corner bead x $____/ft	= $____
	Install ____ ft of expansion joint x $____/ft	= $____
	Texture (SF x $.15 /SF)	= $41.40
	Repair of framing (LF x $____/LF)	= $____
	Install furring @ $____/ft	= $____
	Other:	= $____
Ceiling labor subtotal:		= $172.91
Walls:	Install/tape 443 SF of panels x $.351/SF	= $155.49
	Cut 14 openings @ .1 hrs/cut x $20/hr	= $28.00
	Install ____ ft of corner bead x $____/ft	= $____
	Install ____ ft of expansion joint x $____/ft	= $____
	Texture (SF x $.12 /SF)	= $53.16
	Repair of framing (LF x $____/LF)	= $____
	Install furring @ $____/ft	= $____
	Other:	= $____
Wall labor subtotal:		= $236.65
Room labor subtotal:		= $409.56

Kitchen/dining room labor estimate
Figure 7-34

Room Labor Estimate

Date: 8/6/87 Sam's Sheetrock Page 8 of 9

Job location/description: RESIDENCE AT 4493 CEDAR LN Room: FAMILY ROOM

Estimate for: J.B. SMITH Room dimensions: 21'0" X 22'0" + 8'6" X 8'6"

Labor

Ceiling:	Install/tape 388 SF of panels x $.462/SF	= $179.26
	Cut 4 openings @ .1 hrs/cut x $20/hr	= $8.00
	Install ____ ft of corner bead x $____/ft	= $____
	Install ____ ft of expansion joint x $____/ft	= $____
	Texture (SF x $.15/SF)	= $58.20
	Repair of framing (LF x $____/LF)	= $____
	Install furring @ $____/ft	= $____
	Other:	= $____
Ceiling labor subtotal:		= $245.46
Walls:	Install/tape 777 SF of panels x $.351/SF	= $272.73
	Cut 10 openings @ .1 hrs/cut x $20/hr	= $20.00
	Install 8 ft of corner bead x $.56/ft	= $4.48
	Install ____ ft of expansion joint x $____/ft	= $____
	Texture (SF x $.12/SF)	= $93.24
	Repair of framing (LF x $____/LF)	= $____
	Install furring @ $____/ft	= $____
	Other:	= $____
Wall labor subtotal:		= $390.45
Room labor subtotal:		= $635.91

Family room labor estimate
Figure 7-35

Room Labor Estimate

Date: 8/6/87 Sam's Sheetrock Page 9 of 9

Job location/description: RESIDENCE AT 4493 CEDAR LN Room: LAUNDRY AND PANTRY

Estimate for: J. B. SMITH Room dimensions: 8'6" x 7'6" + 8'6" x 1'0"

Labor

Ceiling: Install/tape 73 SF of panels x $.462/SF = $33.73
Cut 2 openings @ .1 hrs/cut x $20/hr = $4.00
Install ____ ft of corner bead x $____/ft = $____
Install ____ ft of expansion joint x $____/ft = $____
Texture (SF x $.15/SF) = $10.95
Repair of framing (LF x $____/LF) = $____
Install furring @ $____/ft = $____
Other: = $____

Ceiling labor subtotal: = $48.68

Walls: Install/tape 288 SF of panels x $.351/SF = $101.09
Cut 7 openings @ .1 hrs/cut x $20/hr = $14.00
Install ____ ft of corner bead x $____/ft = $____
Install ____ ft of expansion joint x $____/ft = $____
Texture (SF x $.12/SF) = $34.56
Repair of framing (LF x $____/LF) = $____
Install furring @ $____/ft = $____
Other: = $____

Wall labor subtotal: = $149.65

Room labor subtotal: = $198.33

Laundry and pantry labor estimate
Figure 7-36

Estimate Summary

Date: 8/6/87

Page 1 of 1

Job location/description: RESIDENCE AT 4493 CEDAR LN

Estimate for: J. B. SMITH

Materials:

Room: MASTER BEDROOM & CLOSET	$ 158.87
Room: BEDROOM 1 & CLOSET	$ 149.94
Room: BEDROOM 2 & CLOSET	$ 162.90
Room: 2 BATHROOMS	$ 210.46
Room: HALL, FURNACE CLOSET, LINEN C.	$ 117.39
Room: LIVING ROOM & CLOSET	$ 214.75
Room: KITCHEN/DINING ROOM	$ 185.92
Room: FAMILY ROOM	$ 217.97
Room: LAUNDRY & PANTRY	$ 87.64

Materials **total:** $ 1,505.84

Labor:

Room: MASTER BEDROOM & CLOSET	$ 378.81
Room: BEDROOM 1 & CLOSET	$ 377.13
Room: BEDROOM 2 & CLOSET	$ 397.62
Room: 2 BATHROOMS	$ 255.54
Room: HALL, FURNACE CLOSET, LINEN C.	$ 280.48
Room: LIVING ROOM & CLOSET	$ 390.13
Room: KITCHEN/DINING ROOM	$ 409.56
Room: FAMILY ROOM	$ 635.91
Room: LAUNDRY & PANTRY	$ 198.33

Labor **total:** $ 3,323.51

Materials + labor:	$ 4,829.35
Supervision:	$ 500.00
Overhead:	$ 900.00
Contingencies:	$ 170.00
Total cost:	$ 6,399.35
Profit:	$ 670.00
Total estimate:	$ 7,069.35

Sample estimate summary
Figure 7-37

Chapter 8

STARTING YOUR OWN BUSINESS

There are two ways to make a living as a professional drywall installer. You can work for someone else for wages on their payroll, or *you* can be the contractor. Both choices have advantages and disadvantages.

The key advantage to working for someone else is that you leave your work at the job site when the day's work is done. It's not your responsibility to see that the project comes in on time, comes in at or below cost, and leaves a contented customer. That's the contractor's job. The main disadvantage to working for someone else is that your income potential is limited to straight wages or salary. Your chance of increasing your take-home depends on how good you are, how much work there is, and the skill and judgment of the contractor you work for.

If you're the contractor, your responsibilities increase. But so does your income potential. You'll set your own salary. You'll be the one deciding how much work to bid, which jobs to take and when. But being the contractor rather than working for wages is no guarantee that your net income will be higher. In fact, it may be lower, even if you're working harder, longer hours, and carry more of the risk.

Net income means the amount of money that remains after all labor, materials, equipment, overhead and taxes have been paid. Your company will pay taxes on net income. *Gross income* is the money received from customers before paying any bills.

Here are some important considerations when setting up your drywall contracting business. You'll need to make some decisions about the size of your business. You'll want to establish goals, examine the marketplace, look at government regulations, arrange for insurance, locate suppliers, arrange for legal advice and possibly get a loan. In this chapter, I'll discuss each of these considerations and suggest some steps to follow when setting up a profitable drywall contracting business.

STARTING SMALL

When you start any new business, many of your responsibilities will be unfamiliar to you. Mastering these new skills comes easier when you take it in small doses. That gives you the chance to experiment and correct each error before it becomes a major crisis. As the business expands, you'll have a solid foundation of experience to work from.

Starting small will also limit your financial investment in the business, cutting down on your risks. This is particularly important when you're first starting out. You may discover that you don't like working for yourself. You may prefer a steady income rather than going from feast to famine. The business may take more of your time than you want to commit to it. You may not want to accept responsibility for the success or failure of a business. Starting out small is a good way to learn about the business and about yourself as a contractor. Take it one step at a time.

Providing a Financial Cushion

Many small businesses fail every year. And one of the main reasons they fail is that they're *undercapitalized*. You're undercapitalized when you start a business without enough money to cushion you during the lean times. Like other construction businesses, drywall work has its busy times and its slow times. Some months, there will be more work than you can handle. Other times, there may be few jobs to bid on. If you're just starting out, it may be slow for quite a while until you establish a good reputation. You'll need a financial cushion to help pay the bills during those inevitable slow periods.

Limiting the size of your new business can also help you get through the lean times. A smaller business has fewer bills to pay. And reducing expenses reduces the need to find additional work when there's little or none available.

Interest Rates and Inflation

When thinking about the size of your new business, take a careful look at the area you plan to work in. And ask yourself this question: "How much new construction and remodeling work is going on in this area?" Interest rates and inflation both affect the quantity of work available.

When interest rates are low, construction contractors are going to be active. Buyers can qualify for home loans they can afford. There may also be more commercial work available to bid on.

When interest rates are higher, there will be less new construction, but remodelers may be more active. Many remodeling projects are paid for from savings accounts rather than through loans.

It's easier to start a new business when the economy is strong and expanding. But it's also easy to become too large too fast. Rapid growth will increase profits, at least as long as growth continues. But when it stops, as it always will, your overhead can become an oppressive burden. Office rent, loan payments, and administrative payroll go on even when there's little or no work to do. It's easy to end up with less than what you'd have with a smaller operation.

My advice is to resist the temptation to start out in the "megabucks" league. Start smaller, with an eye toward growing in the future at a rate you can handle. Making as much money as quickly as possible may not always be wise. Remember, you want to *stay* in business and stay profitable, not just increase dollar volume. Let your business growth run parallel to your growth as the owner and manager of your business.

Doing Your Own Estimating

If you're fortunate enough to have a business partner who's good at estimating, you can concentrate on other parts of the business. Otherwise, you should probably do the estimating yourself. After all, it's your business. A bad estimate can wipe out your profit and maybe even your business. If you do your own estimating, it's sure to get the care and attention it deserves.

Don't give responsibility for estimating to someone else until you've taken the time to train and supervise that person. Once you know the new estimator estimates like you would, delegate some or all of the responsibility for estimating. But not until then. A careless estimator can be deadly to your business survival.

How Many Employees

Starting out small, you won't need many employees. A two man crew can probably handle most jobs, including the estimating and bookkeeping. In the daytime you'll be installing and finishing drywall. At night and on weekends you'll be estimating, keeping the books, and selling more work.

Once you have more than one job going at a time, you'll need more employees — and at least one supervisor to be sure work is done on time and to your satisfaction. A drywall contractor grossing more than about $150,000 a year will also need clerical help to handle payroll, billing, filing and reports. That person can also answer the phone. The advantage is that someone is usually available to respond to customers during working hours. The disadvantage is that clerical help adds to your overhead, making it harder to submit competitive bids.

As your company grows, so will your need for space. In the beginning, you'll need no more than a desk in a den or bedroom for keeping files, stacking plans, and doing paperwork. For a larger business, a room in a home won't be enough. It's an inconvenience for you, your employees and your family.

Consider Sharing the Load

When deciding how many of your business responsibilities to take on by yourself, consider your own knowledge, interests, and background. If you're installing drywall by day, you may not be up for spending your evenings doing the bookkeeping or other paperwork. There will be bills to be paid, building permits to apply for, materials to order, insurance to arrange and premiums to pay. Someone will have to pay bills, negotiate prices, pay taxes and file payroll reports.

If paperwork like this doesn't interest you, don't despair. You can hire someone to be a bookkeeper. You'll just have to oversee what that person does. But if you're comfortable doing the estimating, paperwork, supervising and keeping the books, maybe you should leave the hanging, taping and texturing to others. There's nothing that says you have to hang the drywall yourself.

ESTABLISHING GOALS

Let's say you're planning to take a trip. First you'll decide where you want to go. You'll find out the distances involved and decide on the best way to get there. If you're driving, you'll choose the route to take and where you want to stay. You'll plan the travel time and the sights you want to see along the way. Setting goals and planning should be part of any major trip. Shouldn't you do the same planning and goal-setting for your career? Long-term goals establish where you're going. Short-term goals set intermediate steps to make sure you're getting there on schedule.

Think of running your business in the same way as you think of making a trip. But this trip will be a journey through time, in a sense. You (and your business) are at a certain point today. You only have three choices of destination. Your business can grow, it can stay the same, or it can shrink and perhaps cease to be.

Let's assume you make the first choice: You want the business to grow. Think about what that means. Over a period of time, the business is to grow in size. That growth is your goal. But you have to define that goal. How big is the business to grow? And over what period of time?

Today is the starting point of the journey. The desired size at the end will be the destination of the journey. This destination is the long-term goal. As with any trip of more than one day, you'll have to select stops along the way. These stops are the short-term goals. You need a goal you can reach in a short period of time. Why? So you know you're headed in the right direction to achieve your long-term goal.

You can't afford to wait until you've reached your long-term goal to discover you took a wrong turn, or you're going too slow. The short-term goals will give you a way to monitor progress and to identify corrections that are needed. Also, short-term goals can give you a feeling of accomplishment. That helps keep you motivated toward your long-term goal.

Let's take an example. Imagine that your business consists of you and one other person. Your gross income is $125,000 per year. In two years, you'd like to be grossing $500,000. Based on the work you did to gross $100,000, you believe you'll need seven people, including yourself. In order to support seven people and reach your gross income goal, you'll have to find more work. It's not practical to try to find that amount of work immediately, or to hire the extra people and then look for the work. You'll have to build up gradually.

Let's assume the work is out there if you are able to expand your work force. You want a competent crew, so you decide to add people gradually and take on more work as they're trained and able to handle it. This may be your first short-term goal: Plan to add another person in the next month and increase your workload enough to keep the three of you busy. Then, based on your results, figure out how fast you think you can and should increase the size of your business. Set other short-term goals so you can reach your long-term goal in the time you have chosen. As time passes, check your progress against your short-term goals and make adjustments as needed.

Believe me, setting goals is essential when you're running your own business. Let's look at some examples of short- and long-term goals you might want to set up in your drywall contracting business.

EXAMINING THE MARKETPLACE

Before you open the doors of your drywall business, examine the marketplace. Determine what the market is for the service you plan to provide. Larger cities are much more diverse than a smaller town or suburb. A large population center will have many contractors who specialize in drywall work. Smaller towns probably have at least a few builders who can hang drywall but don't specialize in that type of work. Since hanging drywall is specialized work, you're probably better off working in or near a larger city where there's a demand for drywall specialists.

But few drywall contractors can do business all over a large metropolitan area. If you're working in a large city, you may want to limit your work area so that your jobs aren't too spread out.

To determine the market for your services in your area, do a *market survey.* Here are the key questions that the survey must answer:

1) How much construction is going on in the area?

2) What types of jobs are available at the present time?

3) Who are the major contracting firms in the area?

4) Which firms are bidding on the jobs that you want?

Talk to general contractors and drywall contractors in your area. Find out all you can about the type of work they do. Once your survey is complete, you should have an accurate feel for the mix of work available in the area and the problems these contractors face.

UNDERSTANDING GOVERNMENT REGULATIONS

The Small Business Association (SBA) can provide you with information on the government regulations that affect your contracting business. There's at least one field office in each state. The local chamber of commerce can give you information that applies to your specific area.

Contractor's License

Your state may require that you have a contractor's license to contract for drywall jobs. The state office of consumer affairs can give you information on how to qualify for a contractor's license. This normally involves taking an examination on the laws, codes and regulations that apply to contracting. A contractor's license will also help you get discounts on materials and may help you get the insurance you need.

Business License

You'll probably need a business license from your city or county to do business in that city or county. If your business is not a corporation, and your full name is not the name of the business, you'll register your business under the *fictitious name law*. The fictitious name you choose for your business can be any name you want, as long as it isn't the same as the name of another organization in your state.

Use Tax and Sales Number

You'll probably also have to get a resale permit from your state. This is required to buy materials without paying sales tax. The law varies from state to state. But usually contractors pay sales tax on the materials they buy. If you're reselling materials, however, you won't pay sales tax, but will have to charge it when the materials are resold. You then remit the tax collected to the appropriate state agency. Part of your bookkeeping task will be to keep records of what you buy and sell.

Employer's ID Number

The Internal Revenue Service (IRS) can provide you with an employer's identification number and detailed information on tax reporting requirements for a business. This tax information is contained in a "Going Into Business Tax Kit." Keeping accurate tax records and making prompt tax payments are part of your bookkeeping.

OSHA Regulations

OSHA stands for Occupational Safety and Health Administration. This is a federal government agency that's part of the U.S. Department of Labor. OSHA regulations are designed to reduce the risk of accident or injury in the workplace. Since you're likely to be hiring employees, these regulations will apply to you.

OSHA regulations change often. Contact OSHA directly for information on the regulations that apply to drywall contracting. You can find the address in your local phone directory white pages under U.S. Government Department of Labor. They have publications available written specifically for small businesses.

Withholding Taxes

When you have employees, you'll have to withhold taxes from the employees' wages. The IRS can provide you with the specific details and publications that explain federal withholding requirements. In most states, you'll also be required to withhold state income taxes from your employees' wages.

ARRANGING FOR INSURANCE

Insurance is essential when running your own business. Finding a competent insurance broker is important. Your broker should be familiar with the minimum insurance coverage required for an employer and his employees. A good broker will also be able to recommend any special coverages you'll need. The minimum insurance required by law may not be enough to fully protect you or your workers. A good broker can work out a plan that provides a prudent amount of insurance, plus plans for additional coverage as you expand.

Here are the types of insurance you'll want to consider:

- You'll want to protect your investment in physical property against loss and damage.

- Your state will have specific requirements for Worker's Compensation coverage.

- You'll want to carry suitable liability insurance to cover any accidents that might occur. Liability insurance should cover damage to property and injury to your workers and to nonemployees.

- You may also want to carry life insurance.

- For some jobs you'll need performance bonds to qualify to bid on the work.

The insurance broker you select should be able to quote costs for these coverages and recommend an appropriate insurance package. The agent should be able to back up those recommendations with facts so you can select the insurance coverage that best meets your needs.

LOCATING SUPPLIERS

It's important that you find *reliable* suppliers for the materials and equipment you'll be using in your business. Start your search for suppliers with the phone book and *Thomas Register*, which is available at most local libraries. It has listings of manufacturers by type of product. Select your favorite supplier by collecting some information:

1) What credit terms are they willing to give to a new account?

2) What discounts can you expect?

3) What's the lead time on standard items? On special-order items?

4) Do they have most drywall construction materials in stock, or do they have to special-order them? Your local supplier may stock 1/2-inch standard and water-resistant drywall 4 x 8 panels. But 12-foot long panels may be a special order item.

5) What quantity of special-order material is required before the supplier will place an order? If you need less than that quantity, there may be no practical way to get what you need. Instead, you'll have to install shorter panels and allow extra labor in your estimate.

6) Does the supplier deliver on time? This is important. The most competitive bid is worthless if your supplier can't get the materials to you on time. Ask for references, the names of contractors who buy from that supplier.

LEGAL ADVICE

To find a good lawyer, ask your banker, friends and other contractors for a recommendation. When you need legal assistance, interview several prospects. Find someone who will take a personal interest in your business and represent you fairly and aggressively at moderate cost.

Your lawyer should be able to recommend a form of organization: sole proprietorship, partnership or corporation. He should also be able to draft a good contract form that will protect you and your clients.

Sole Proprietorship

If you're going to be the only owner, you'll probably operate as a *sole proprietorship*. That's a business owned and operated by one person. It's the easiest and least expensive type of business to start.

As a sole proprietor, all the profits show up as income on your tax return. You have total control over the operation. You can make changes quickly to respond to changing situations.

Of course there are disadvantages to balance these advantages. You'll be liable for the full amount of your business debts. And this liability extends to all of your personal possessions. If you can't work, you'll almost certainly hurt the business. And as a sole proprietor, it may be harder to accumulate working capital.

Your lawyer may recommend another type of ownership, based on your financial situation, the type of business, the number of employees, the risks and your tax situation. He may recommend either a partnership or a corporation.

Partnership

A *partnership* is the association of two or more people as co-owners of a business. The rights and obligations of each partner are set out in the partnership agreement, which is like a contract between the partners.

The advantage of a partnership is that it can bring the skills, knowledge and resources of several people into one business. It's relatively easy to form, and can be flexible enough to meet most business situations. It's relatively free from special taxes and governmental control. Business profits flow through to each partner every year. The partnership files only an information tax return, paying no taxes itself.

There are disadvantages. At least one partner has unlimited liability. The partnership is dissolved when any partner dies. The business can be bound by the acts of just one of the partners. And it may be hard to buy out a partner.

The partnership agreement should anticipate any problems the partnership will face. For example, if there's a dispute between the partners that can't be resolved, some provision should allow one partner to buy out the other at an agreed price. The agreement should also make it clear what happens if one partner dies or can't work.

Corporation

A corporation is more complex than a partnership. It's a separate legal being and files a tax return just like individuals and families do. It's formed by authority of state government. A corporation that does business in more than one state must comply with laws in each state where it does business.

Here are the advantages of a corporation. The owners' or stockholders' liability is limited to the amount invested. Ownership is easily transferred by selling stock. A corporation can go on forever, even after a stockholder dies.

And now the disadvantages: A corporation's activities are limited by its charter and laws. Minority stockholders have little voice in how the company is run. It's more expensive to set up a corporation than a partnership or sole proprietorship. The taxes are more complex.

One type of corporation you may be interested in is the Subchapter S Corporation. This type of corporation treats the corporation's profits or losses as ordinary income or loss to the individual stockholder. It's intended for smaller businesses. The losses on the stock of a small business corporation can be treated as deductions against ordinary income. Your legal counsel will recommend forming a Subchapter S Corporation if that's appropriate.

Setting Up Contracts

Many drywall contractors rely on verbal agreements. But verbal commitments can be hazardous to your financial health. My advice is to get a written agreement on anything but the smallest jobs. Don't let faulty memories cost you money. Have your legal counsel prepare a standard agreement for you or use one of the preprinted contracts you can buy in any office supply store. Figure 8-1 shows a typical contract that can be used by almost any construction company. Figure 8-2 is another standard form, an additional work authorization for change orders. You can customize these forms by having your company name and address printed on them.

Proposal Page No. ____ of ____ Pages

Your Construction Co.

Your address

Phone 254-0000 Contr. Lic. No. 123456

PROPOSAL SUBMITTED TO		PHONE	DATE
STREET		JOB NAME	
CITY, STATE AND ZIP CODE		JOB LOCATION	
ARCHITECT	DATE OF PLANS		JOB PHONE

We hereby submit specifications and estimates for:

We Propose hereby to furnish material and labor — complete in accordance with above specifications, for the sum of:

______________________________ dollars ($ ____________).

Payment to be made as follows:

All material is guaranteed to be as specified. All work to be completed in a workmanlike manner according to standard practices. Any alteration or deviation from above specifications involving extra costs will be executed only upon written orders, and will become an extra charge over and above the estimate. All agreements contingent upon strikes, accidents or delays beyond our control. Owner to carry fire, tornado and other necessary insurance. Our workers are fully covered by Workmen's Compensation Insurance.

Authorized Signature ____________________

Note: This proposal may be withdrawn by us if not accepted within ____________ days.

Acceptance of Proposal — The above prices, specifications and conditions are satisfactory and are hereby accepted. You are authorized to do the work as specified. Payment will be made as outlined above.

Signature ____________________

Date of Acceptance: ____________________

Signature ____________________

FORM 118 COPYRIGHT 1960 - NEW ENGLAND BUSINESS SERVICE, INC. GROTON, MASS. 01450

Typical contract
Figure 8-1

Another document you should have to help close sales is a general fact sheet about your company. This sheet should tell the client who to contact in the case of questions and how to contact them. The fact sheet can also tell the client how you will

ADDITIONAL WORK AUTHORIZATION

Your Construction Co.
Your address
Your phone number

OWNER'S NAME		PHONE	DATE
STREET		JOB NAME	JOB NUMBER
CITY	STATE	STREET	
EXISTING CONTRACT NO.	DATE OF EXISTING CONTRACT	CITY	STATE

You are hereby authorized to perform the following specifically described additional work:

ADDITIONAL CHARGE FOR ABOVE WORK IS: $ ________

Payment will be made as follows: ________

Above additional work to be performed under same conditions as specified in original contract unless otherwise stipulated.

Date ________ 19 ____ Authorizing Signature ________ (OWNER SIGNS HERE)

We hereby agree to furnish labor and materials – complete in accordance with the above specifications, at above stated price.

Authorized Signature ________ (CONTRACTOR SIGNS HERE) Date ________ 19 ____

THIS IS CHANGE ORDER NO. ________

NOTE: This Revision becomes part of, and in conformance with, the existing contract.

Typical change order
Figure 8-2

proceed with the work, outlining the steps you'll take during the course of the job. This will give the client an understanding of what you're doing and what you have left to do. A good fact sheet will answer most of the routine questions you'll receive.

GETTING A LOAN

Your new drywall company will need some working capital. If you don't have savings available, you'll have to find a short-term loan, longer term loan or equity capital. Your local bank and the Small Business Administration will be good places to start. Let's look at these three types of financing.

Short-Term Loans

Short-term loans are limited to a few days, weeks or months. You repay a short-term loan by converting assets into cash. You can use short-term loans to finance accounts receivable (money owed to you) for 30 or 60 days. Or you can use them to carry inventory or finance equipment for a longer period of time, maybe even six months.

Lenders usually expect these loans to be paid back once the loan has served its purpose. If you borrowed the money to cover your accounts receivable, you'll be expected to repay the loan once your customers have paid you. If you took out the loan to purchase inventory, the lender will expect his money back as the inventory is converted into completed work. For example, if you borrowed the money to buy drywall materials for a job, you should pay the lender back when your customer pays you for the work you did.

Short-term loans are often unsecured. To get an unsecured loan, you'll need a good reputation for paying creditors. If you haven't yet built up a credit history, you'll need to borrow against owned assets. If you can't repay the loan, the lender has the right to claim the assets you pledged to secure the loan.

There's one special type of short-term loan that you should know about. A *line of credit* is a prearranged loan made with your bank. It allows you to draw up to the maximum approved amount of money for a short period of time, such as 30 days. It's very handy for material purchases and other expenditures that will be paid off in the near future.

Term Loan

Term loans offer a longer period of time for repayment. You may have from one to five years to repay the loan. A term loan is a secured loan — you'll have to provide collateral to get it.

The key advantages to term loans are that you can get a larger amount of money, and the loan repayment is spread over a longer period. That reduces the size of each payment.

Equity Capital

Equity capital isn't a loan that you'll be paying back. Equity capital is money you get by selling an interest in your business. Your business becomes an investment for the person who provides the money. An investor is probably looking for income and an increase in the value of the investment. Relatives and friends are good sources of equity capital. Anyone who believes in you and wants to see you succeed may be willing to invest equity capital in your business.

Local Banks

Discuss credit and loans with your banker. Find out what their requirements are for extending credit to businesses like yours. Usually your banker will want you to:

- Show that you deserve a good credit rating.
- Put up at last half of the money needed to start your business.
- Show that you have business experience.
- Provide a complete description of the business, including a financial statement.
- List collateral you intend to use to secure the loan.
- Show exactly how you plan to use the loan money if the loan is granted.
- If you've been in business for a while, you'll have to show the profits you've made over some recent period of time, such as the last three years. Usually your banker will want a copy of your income tax return for the last two years.

Local banks often put restrictions on loans to protect themselves against poor management practices by the borrower. The longer the loan repayment period, the greater the number of restrictions placed on the loan. Examples of loan restrictions include the specific repayment terms and periodic reporting on the condition of the business.

You'll find these restrictions in the section of the loan titled "Covenants." *Negative covenants* are things the borrower can't do without approval from the lender. Negative covenants are designed to protect the lender against action by the borrower that would put the loan collateral or repayment in jeopardy. *Positive covenants* define what the borrower must do. You can negotiate these covenants. Work out the loan agreement terms before you sign the loan agreement. Try to get terms you know you can live with.

The SBA

The Small Business Administration guarantees loans to small businesses. The SBA also assists small businesses in getting government contracts. Terms of SBA loans vary according to the type of loan. These loans fall into two categories:

- The SBA will guarantee loans made to small businesses by private (nongovernmental) lenders up to a maximum of $500,000.

- The SBA has authority to make loans directly to small businesses. These direct loans are limited to $150,000. They will do this only when a small business can't get a loan from a private lender or can't get an SBA guarantee for a private loan.

The SBA is more likely to guarantee a loan made by a private lender, rather than make a loan directly to a small business. This is because the SBA has limited resources and funds for direct loans. And by law, the SBA cannot guarantee or make a loan unless the small business can't get a loan elsewhere at reasonable rates.

Requirements for SBA loans include:

1) The business must be independently owned and operated for a profit.

2) It must *not* be the dominant business in its field and it must meet certain size standards.

3) Certain types of businesses aren't eligible.

4) The applicant must meet certain standards.

5) The loan must be secured with some form of collateral.

Providing Collateral

Depending on the size of the loan and your credit history, you may have to provide some form of collateral. In some cases, a guarantee by a third party may be all that's needed. For example, a guarantee by the SBA will be acceptable by many lenders. Collateral could be the assignment of a lease, real estate, savings accounts, stocks or bonds or some piece of equipment such as a truck.

To determine if your collateral is acceptable, the lender considers the value of the collateral compared to the amount of the loan, evaluates the likelihood that you'll be able to pay back the loan, and reviews the salability of the collateral. You can negotiate with your lender the type and amount of collateral required.

When a third party *endorses* your loan, they're offering to be liable for payment of the loan in case you don't pay it. Having someone endorse your loan may eliminate the need for collateral. This is especially true when you're just starting out, or when your credit reputation isn't solid enough to warrant the size of loan you require. The endorser of your loan may be asked to pledge collateral in addition to your own collateral.

Someone who *co-makes* a loan with you is equally liable for payment of the loan. The lender can collect the loan directly from you or from the co-maker.

Someone who *guarantees* your loan by signing a guaranty commitment is known as a *guarantor*. A manufacturer can be a guarantor for a customer. An officer of a corporation can be the guarantor of a corporate loan.

You'll have to decide whether your company will extend credit to customers. If you decide to do so at some point, limit the amount and set specific repayment terms.

That's about all I have to say on starting your business. The next chapter should help you fit the final piece into the puzzle: managing the business you finally got started.

Chapter 9

ORGANIZING FOR PROFIT AND EFFICIENCY

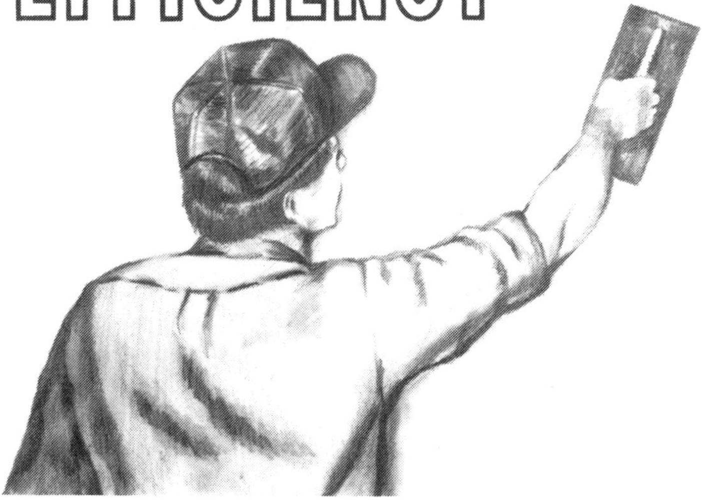

Running a successful drywall contracting business requires more than just keeping track of the money coming in and the money going out. Your company has to be organized to operate efficiently and profitably. That's not always easy. This chapter will lay out some of the basic principles you should know and suggest ways to overcome the most common drywall contracting problems.

To be competitive and successful in drywall contracting you have to hire and retain a balanced work force, motivate your employees to give their best performance, schedule your jobs so they finish on time, manage your own time and paperwork efficiently, and sell enough of the most profitable jobs to stay busy, even when work is scarce. We'll look at each of these problem areas one at a time, beginning with selecting employees.

HIRING A BALANCED WORK FORCE

Hiring a balanced work force means hiring the right number of people to handle the work load. It also means matching the worker to the right task and responsibility. Let's look at some important points to remember as you hire employees and assign work.

Hiring the Right Number of People

As your business grows, so will the number of employees on your payroll. The first employee will be a helper and assistant to apply drywall. A company that has more than one work crew will need a supervisor and, eventually, clerical staff to handle accounting, billing, taxes and payroll.

As a business grows, it can afford more overhead expense without decreasing profits. But there's a delicate balance between having the right number of people to handle the work load and having an overhead that's too high. Excessive overhead is the surest way to reduce or eliminate profits. As long as everyone is busy and productive, the company is operating efficiently and overhead is probably no higher than necessary. But staying busy isn't easy in the drywall contracting business.

Once you've been in business for a while, you'll be able to accurately forecast incoming work and the number of employees required to handle it. Until you develop that skill, it's hard to commit to hiring additional employees when you're not certain how long you'll be able to keep them employed. Here's how experienced drywall contractors forecast workloads and adjust payroll accordingly.

1) Hire temporary office and craftspeople for specific tasks on an as-needed basis.

2) Use an answering machine or an answering service during the day, instead of hiring a full-time receptionist. But be sure to check for incoming messages several times a day. Many answering services will contact you with a beeper when there's an important message. If you spend several hours a day in your car, consider investing in a car phone.

3) Do your own paperwork. If you're starting out as a two-person operation, a lot of the installation work can be handled by one person. This leaves the other person free to do the estimating, bidding, billing and other paperwork.

4) If you're doing more than about $10,000 gross a month, consider having an accounting service handle your accounting, billing, taxes and payroll. Many will be listed in the Yellow Pages of your phone book. Some specialize in construction contractors. They do the paperwork, but you still have to provide the information needed to prepare accounting records, create invoices and statements, generate the payroll and payroll tax reports, and file tax returns.

Matching the Worker to the Task

Everyone supervising employees has the problem of matching the worker to the task or job responsibility. It would be great if you could have all experienced, top-notch people working for you. But my experience is that most drywall companies have a mixed bag — people with different skill levels and different levels of productivity.

Here's what a well-balanced work force might look like. You'll have one or two relatively new people who are in training. You'll have some people with average skill levels and productivity. And you'll have some very experienced and relatively fast workers. There may also be people who are fast but not very thorough. You may have others who are very thorough but not very fast.

It's obvious that you can't assign all new people to a job. The schedule and quality of the work would suffer. Too much supervision would be required. New people benefit from working with those who have more experience. You also can't afford to load up a job with only your most experienced people. Then you wouldn't be able to take on any new work. Balancing the work assignments helps you keep costs moderate while ensuring quality work.

Crew leaders must be just that: leaders. Work experience alone isn't enough. A crew leader should accept responsibility for seeing that the work is done according to your standards and on time. You'll find that some very experienced workers don't want the responsibility of leading people. They're most content when they can work with a minimum of supervision and work either alone or with one other person. There's no advantage to making supervisors out of these people. They're most productive when allowed to work under the conditions they prefer.

I've known workers who wanted to be leaders but who didn't have the skills that supervisors require. You may be able to train these people to become effective leaders. But this will require patience, guidance and close supervision on your part. Start out with small supervisory tasks and see how they handle them. By guiding them, making corrections and gradually increasing their level of responsibility, you may be able to make effective leaders of them. Avoid giving anyone too much responsibility too fast. Effective leaders can become totally ineffective when given too much responsibility too soon. Advance people at a pace appropriate to their skills and talents.

MANAGING YOUR EMPLOYEES

You're the boss. It's your responsibility to monitor the strengths and weaknesses of your employees. This includes the speed and quality of their work, their reliability and ability to handle responsibility. It's to your advantage and theirs to find ways to make each a more productive, more reliable employee.

Some employees will make good foremen. Others will excel when they have no leadership responsibilities. Every employee is unique. It's up to you to motivate each employee so you get the best performance possible. Recognize good performance, be fair and impartial with your employees, and insist on an honest day's work for a day's pay.

The New Employee

It's important to get the new employee off to a good start. Every new employee needs a clear understanding of the work you expect him to do and how you expect him to do it. He needs to know his role, the tasks involved, and the standards of quality and performance you expect. And he needs to be rewarded when he performs in accordance with your expectations.

Prepare a job description. Include in the job description as many details of the job as possible. Review the job description with each new employee. As you review it together, ask the employee questions about job requirements until you're satisfied that the employee understands the work as you've described it.

For a foreman's job description, for example, you might include "working with the client." This will include personal contact, from advising the client on the work schedule to arriving at an agreement for a change in the type of texture to be used. You can't go over each task that might come up. But you can help an employee understand the general scope of his responsibilities.

Once the employee clearly understands the work involved, let him know the standards of quality and performance that you expect. Understanding the required tasks and necessary standards will help any employee do his or her best for you. Monitor that employee's work to observe errors. Suggest improvements when you note errors. When performance needs to be improved, point this out. The best example is the example you set yourself. This is called leading by example. If an employee is doing a good job, be the first to offer a pat on the back. A compliment may inspire even better work. If good work goes unnoticed, there's little incentive to improve. When errors go undetected, they're likely to be repeated.

Keeping employees advised of their performance will help them grow and take on added responsibilities. When they know you're interested in their performance, they'll do better work. When I was still new in the business, I had a boss who told me, "If you don't hear anything, you're doing O.K." That's negative feedback. The only communication is criticism. It destroys employee incentive.

As your company grows, you'll need key people you can depend on to take responsibility for important parts of the business. You won't be able to do everything yourself. A larger business can only be successful when each person carries his or her part of the load. It's up to you to develop people who can do this.

Reward Good Performance

When someone does well consistently or assumes responsibilities that are greater than the position demands, reward them. You can do this verbally or with a written commendation. Or you can give them a raise or a bonus. The type of reward you give will depend on the level of performance, the employee's current salary and the company's ability to pay more. But don't overlook the importance of rewarding good performance, even if you can't offer a pay raise. If you reward initiative and superior work, you'll likely get more of it. If you ignore it, it may disappear.

Don't let your supervisors take all of the credit for work well done. Good performance is usually at least partly the result of good leadership. But those who are led deserve their share of the praise and rewards. Recognize and reward individual contributors who do outstanding work.

Stress Teamwork

Any job can be done better when people work together toward a common goal. There must be some leaders, but they don't have to be heroes. Recognizing only the leader of a team will undermine the teamwork effort.

People can only work well together, and work toward a common goal, when they understand the entire operation and the role of each team member. There must be good communication among the members of the team. It's up to the team leader to establish a good communication network among team members. When a team is working together, the work gets done faster and the quality is better.

Set Up a Suggestion Plan

When your company is small, it's easy to talk to each of your employees in person. Good two-way communication is simple. Everyone benefits when ideas and suggestions are traded back and forth.

As your company grows, informal communication channels may break down. You may not be able to talk individually to every employee. And since the people doing the work are in the best position to see ways to improve their own jobs, you may miss out on some important suggestions. Encourage employees to talk with you informally. Make yourself available after work or at quitting time to discuss anything any employee has on his or her mind. If that doesn't work, actively solicit suggestions. Seek out the advice of employees, anyone from your oldest employee to the most recent hire.

Carefully evaluate the suggestions that are made. You won't be able to implement all of them. There might be a suggestion that would make one person's job easier at the expense of another. But there'll be some suggestions that benefit the company and everyone in it. The gain may be in the form of better working conditions, more jobs, more job satisfaction or greater profits. Reward those who make successful suggestions. This will be an incentive for others to put their minds to work to improve things. The reward doesn't have to be a pay raise. But it should be money when the suggestion improves company profits. Take advantage of all of the minds you have working for you. The more people who think, the more good ideas you'll have to choose from.

Profit Sharing

Consider a profit sharing plan. People who have a vested interest in the company will take a greater interest in seeing that things are done right. This is because the better they do their work, the more they have to gain. Profit sharing doesn't have to be part of the package offered to each new employee. It can begin after the employee has been with you for a fixed period of time. Every profitable corporation should have pension and profit sharing plans for key employees. The Yellow Pages of your phone book lists names of companies that can set up and administer a plan for you. Look under "Pension and Profit Sharing Plans."

Delegating Responsibility

The larger your company becomes, the more important good organization becomes. Growth usually means the boss has to delegate more responsibilities to others. Here are some steps to take when you begin delegating responsibility:

1) Make a list of the tasks you want to delegate.

2) List the people who are candidates to take over these tasks.

3) Match names to tasks, based on each person's qualifications.

4) Some tasks will need daily attention. Others will need less frequent attention. Weigh the time needed for each task. Review your list to make sure no one is overloaded. You may have to

do some rearranging of work assignments to even out the work load.

5) Once you have your list of names and tasks, make up an organization chart. This chart should show the responsibilities of each person and indicate the lines of authority.

6) Review the organization chart with each person to make sure he or she understands the responsibilities assigned.

7) Post the organization chart where everyone can refer to it.

8) As the work load and personnel change, revise your organization chart to match the current situation. A good organization chart can prevent problems. It helps everyone understand the assigned responsibilities.

JOB SCHEDULING

Scheduling is simply planning work in a logical sequence so the job is finished on time. There are four key steps to successful scheduling:

1) *Break the job down into tasks.*

2) *Figure the amount of time required to perform each task.*

3) *Figure the number of people needed for each task.*

4) *Draw a bar chart showing the start date, time needed for each task and the required completion date.*

Scheduling doesn't require an elaborate display board or an investment in a computer program. A few notes on the back of an envelope may be enough. On small jobs, scheduling may require only a few mental notes. What you do isn't as important as doing *something.* Have a schedule on every job. Know when you want to start work. Plan ahead so the materials are there on time, the crew is there when needed, the tools and supplies are available, and know when the job should be finished. Those are the minimums. If doing more scheduling than that will improve profits, fine. Do it. But doing less than the minimum will cause problems, even on the smallest job.

Let's look at a sample project. Say we're installing and finishing drywall for one room. Assume the framing doesn't require any straightening. Here's how to schedule the project. First we'll break the job down into tasks. Then we'll add the time required for each task and the number of people. Our breakdown looks like this:

Task	Labor hours	Number of workers
Install wallboard	8	2
Tape joints and spot fastener heads	12	2
Add 2nd coat to joints and fasteners	9	1
Finish joints	6	1
Texture	8	1

Next we can draw our bar chart showing the start date, the time needed for each task, and the required completion date. See Figure 9-1.

```
        I         T         2         F         Tx
Start o---------o---------o---------o---------o---------o Completion
```

Legend: I = Install panels
T = Tape joints and spot fastener heads
2 = 2nd coat of joint compound
F = Finish
Tx = Texture

Bar chart
Figure 9-1

The bar chart will show whether you've allowed enough time between the start date and finish date to complete the work. It also shows whether your finish time will be earlier or later than the required completion date. The difference between the *projected* finish date and the *required* finish date is called *float.* If your chart shows that you'll finish earlier than the required finish date, there's a *positive float.* If it shows that you'll finish later than the required finish date, there's a *negative float.*

Positive Float

When there's a positive float, you'll have extra time available to take on other tasks, at least in theory. Of course, this assumes that your schedule is accurate and that there aren't any unexpected delays. Delays can occur because someone else missed a deadline. For example, your material dealer didn't deliver drywall on time. Or they can occur because something just doesn't go right. You've heard the old saying, "If something can go wrong, it will." Always allow extra time in your schedule for the unexpected delay. This time is known as *contingency time.* As experience will show you, it's a rare job where no contingencies occur.

The amount of contingency time you allow will depend on the following conditions:

- The size and complexity of the job. There are more opportunities for things to go wrong on a big job. A complex construction task, such as a multilevel ceiling, increases the chance of having problems.

- The experience of your work crew. An experienced crew will have less problems. Delays won't disappear, of course, but you can expect fewer of them.

- The weather. With higher humidity or colder temperatures, compounds will take longer to dry and adhesives will take longer to set.

- The size and location of your material supplier. If your supplier isn't large enough to keep a good stock of materials, you may not be able to get all materials when you need them.

Negative Float

A common reason for negative float is that you can't start the job when you're supposed to. The framing might not be finished, the roof might not be loaded yet, the wiring might not be complete or the plumbing might not be installed. If you can't start on time, there's a good chance you won't finish on time either. That's the most common reason for a negative float. But your bar chart can show a negative float, even without delays in starting the job.

When you have a negative float, let your customer know about it. Maybe you can negotiate a revised completion date. This can be based on a new start date, if your work is being held up by someone else. Or you can set up a new total elapsed time from start to finish. You may have to schedule overtime or a larger crew, or both, to make up for lost time. Keep your client advised of problems that affect your schedule. Work out mutually acceptable solutions. That will protect your reputation as a professional, reliable contractor.

No client wants to hear about delays. But if you wait until the last minute to let him know, it makes the problem worse. If the first time he hears about the delay is when you miss your completion date, the first impression is that you're incompetent. And it won't matter to him if someone else caused the delay. A bad reputation spreads faster than chicken pox through an elementary school. Owners and general contractors talk to other owners and contractors. Protect your good reputation. It's your most valuable asset.

Scheduling Multiple Projects

On larger jobs, I recommend drawing a bar chart to show the elapsed time for each task. This is especially important if you have several jobs going at once, if your business is growing, and if crew time or schedules are tight. But don't wait until the fat's in the fire to start perfecting your scheduling skills. Start with a smaller job when a mistake or two won't provoke a crisis. You'll need to use bar charts when your business gets larger. And the more practice you have setting them up, the easier it will be later on.

As your business grows, you'll be dealing with more than one job at a time. And the more jobs you have going at once, the more important good planning becomes. Bar chart schedules let you anticipate problems and avoid delays. Figure 9-2 shows a bar chart schedule for three jobs that overlap in time.

After you've figured out the time required for each task, determine the number of people you need to do the work.

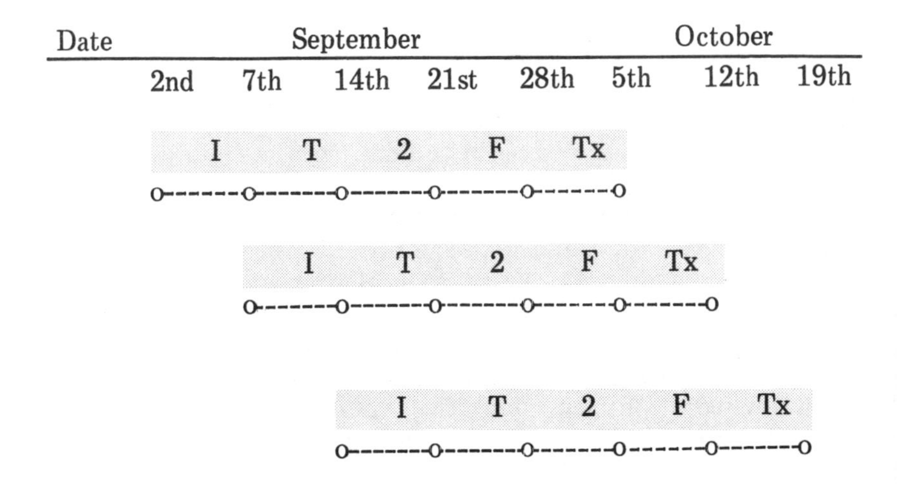

Legend:
I = Install panels
T = Tape joints and spot fastener heads
2 = 2nd coat of joint compound
F = Finish
Tx = Texture

Bar chart schedule for three jobs
Figure 9-2

Include this in your bar chart, as shown in Figure 9-3. Total the number of people you need each week to cover the labor required for all of your projects.

Figure 9-4 shows how to mark progress on your bar chart. As each task is completed, just change the broken line to a solid line. At a glance, you can see what's done and what remains to be done. Figure 9-4 shows us that the third job is behind schedule, while the second job is ahead of schedule.

For any schedule to be of value, it must be kept up to date. You can update the schedule daily, weekly or at some other interval, depending on the length of the schedule and the size of the project. If you're working on a three-month project, there probably isn't much value in updating the schedule on a daily basis. But if you're working on a three-day project, a daily update may be essential. Otherwise, you may not discover a problem until it's too late. The crew is already on site waiting for the next load of drywall!

Date:	September					October		
	2nd	7th	14th	21st	28th	5th	12th	19th
	I	T	2	F	Tx			
	2	2	1	1	1			
		I	T	2	F	Tx		
		2	2	2	2	2		
			I	T	2	F	Tx	
			2	1	1	1	1	
Total workers:	2	4	5	4	4	3	1	

Legend:
I = Install panels
T = Tape joints and spot fastener heads
2 = 2nd coat of joint compound
F = Finish
Tx = Texture

Bar chart schedule showing number of workers required
Figure 9-3

MANAGING YOUR TIME

As the owner and manager of a business, there will be many demands on your time. Your business can take *all* of your time if you let it. But it shouldn't. If it is, you're doing something wrong. You're also making a mistake if you don't devote *enough* time to running your company. How much time is enough time? This depends on you, your company, and how well you manage your time.

Date:	September			Current date		October		
	2nd	7th	14th	21st	28th	5th	12th	19th
		I	T	2	F	Tx		
		2	2	1	1	1		
			I	T	2	F	Tx	
			2	2	2	2	2	
				I	T	2	F	Tx
				2	1	1	1	1
Total workers:		2	4	5	4	4	3	1

Legend: I = Install panels
T = Tape joints and spot fastener heads
2 = 2nd coat of joint compound
F = Finish
Tx = Texture

Bar chart showing actual progress
Figure 9-4

Make Every Minute Count

The key to managing your time efficiently is to *make every minute count*. You shouldn't have to spend 12 hours doing 8 hours' work. Here are four keys to using your time productively:

- Make a list of what you want to get done. Next to each item on the list, write in the amount of time it should take to do it. Leave some extra time for unexpected problems.

- You'll probably discover there's not enough time to complete some of the items on the list. When you've used up the allotted time for an item, stop working on it. Go on to the next item. You can carry the unfinished item over to tomorrow's list.

- Stick to your list. Don't get sidetracked. Following the list lets you focus energy on one item at a time. You may have

43 things you need to get done, but you can only do one thing at a time. Treat each item as the only thing you have to do while you're working on it. If you're working on one item and start thinking about other items at the same time, it cuts down on your productivity.

- Don't let other people schedule your time for you. You're bound to get phone calls and interruptions while you're busy working on one of the items on your list. These phone calls and interruptions are part of your work. You can't ignore them. But you can limit the time of each call. Get right to the point of the call and end it as soon as possible. If the call isn't urgent, and you're in the middle of an important task, ask the caller to call back later.

  Handle interruptions the same way. If an employee needs assistance with a problem, here's how you can handle the situation. First decide what's urgent and what isn't. Decide whether the problem can be solved at the moment or will take additional time. If the problem is going to take time you don't have at the moment, and if it doesn't have to be solved at that instant, set a time to deal with the problem later. Encourage people to do all they can to resolve problems themselves. Ask them to recommend a solution for you. This encourages people to develop their own problem-solving skills.

Conducting Effective Meetings

Meetings are nonproductive in the sense that they don't directly increase your income. But sometimes they're necessary. They're often the most effective way to gather or give information or to reach a decision or agreement. Make your meetings work *for* you instead of against you. Here are eight keys to conducting productive meetings.

1) *Don't hold a meeting on Monday morning or Friday afternoon.* On Monday mornings, people are readjusting to work after being on a weekend schedule. Their minds may not be performing up to speed on work matters. They're trying to orient themselves to the tasks left over from the previous week and to the new tasks to be started. On Friday afternoons, people may be thinking about the upcoming weekend activities. They'll be tired from the week's work. If you assign important

items to them on Friday afternoon, they may forget them by Monday morning. Try to avoid Friday afternoons and Monday mornings as meeting times whenever possible.

2) *Do keep meetings short.* Attention wanders after the first hour, even for the most enthusiastic employee. Most topics don't need more than a few minutes of discussion. If there are so many items that more than an hour is necessary, set up another meeting to take care of it. Don't try to cover it all in one meeting.

3) *Use an agenda for your meetings.* Meetings that last longer than an hour usually do so because time is being wasted during the meeting. Detail the specific topics you need to cover in the meeting. Stick to that agenda. You're calling the meeting for a specific purpose. The agenda is the road map from where you are to where you want to be. If you don't stick to the agenda, there's a good chance you'll never reach the goal.

4) *Appoint someone to control the meeting.* There's often someone in the meeting who wants a forum to discuss unrelated topics. This person will go off on tangents, if the person presiding permits it. Whoever is in charge of the meeting will have to control discussion. Follow the agenda.

5) *Don't invite people to a meeting unless they are necessary participants.* People who aren't needed in the meeting are wasting their time and could be doing something else more productive. And people who aren't needed are more likely to disrupt the meeting. They may not be doing it on purpose, but they can easily get the meeting off track, making it less productive.

6) *Don't schedule regular meetings.* Regular meetings tend to degenerate into nonproductive get-togethers. When you have a need for a meeting, schedule one. That first meeting may show a need for a follow-up meeting to discuss the results of actions planned at the first meeting. It may seem like you should continue having these meetings on a regular basis. But you'll find that regularly scheduled meetings get less and less done. Worse, they can become a social gathering rather than a business meeting.

If you run into a situation where a series of regularly scheduled meetings are a genuine necessity, keep them short and to the point. You may be in the middle of a complex job and need regular meetings for a while to keep everyone working together. Invite only the necessary participants. And use the meetings only for the purpose intended. Discontinue them as soon as possible. Have a clock in the meeting room. Announce the purpose of the meeting and the projected length of the meeting when you start.

7) *Make it clear ahead of time to all meeting participants that you expect them to come fully prepared to discuss the items on the agenda that relate to them.* That's only fair. That's the only way to be sure everyone comes prepared to participate.

8) *Appoint someone to be responsible for taking notes and distributing copies of them after the meeting.* Once the meeting is over, it's hard for the participants to remember all the details of the conclusions reached at the meeting. The minutes serve as a handy reference. They jog people's memories and help ensure that things get done as agreed upon in the meeting.

Follow these guidelines, and your meetings will be shorter, less frequent and more productive.

MANAGING PAPERWORK

The more construction projects you take on, the more paperwork you'll have to do. If you're not careful, you can end up spending more time shuffling paper than doing productive work.

Your records, orders, billing, estimates, schedules and contracts all require paperwork. Here are three suggestions for making your paperwork less of a burden. First, minimize the amount of paper you handle. Second, minimize the amount of time you spend on each piece of paper. Third, set up files that make important papers readily available. A personal computer can also help you manage your paperwork. The less time you have to spend on paperwork, the more time you'll have for other things.

Minimizing the Amount of Paperwork

Cut down on the amount of paper passing through your hands by delegating some of the paperwork details to the people who work for you. As the owner and manager, you'll need to stay informed about day-to-day operations. But you can get this information by looking at the schedule, checking work assignments and checking estimated costs against actual expenses. You don't need to read every line of print on every piece of paperwork that's created. Let's look at a sample situation.

Assume you're the owner and manager of a moderate-size contracting firm with a good organization chart and responsibilities delegated to several employees. There are three people working in your office for at least part of the day. They handle customer inquiries, billing, filing, ordering materials, some of the estimating, accounting and payroll and most of the paperwork. You've assigned supervisors the day-to-day responsibility for the operation of various jobs, including the paperwork relating to their operations. You retain the responsibility for overall management, but you don't need to know about all the details on each job. Your supervisors report to you regularly, usually at the end of each day. This keeps you posted on job progress and abreast of problems — without generating any paperwork. But that's just the beginning.

Minimizing the Time Spent on Paperwork

Once you've cut down on the amount of paper you have to handle, the next step is to handle that paper as efficiently as possible. Handle each paper only once. After you read it, you can do several things with it. You can throw it away if it has no further use. You can route it to others in your company if the paperwork contains information they should have. Or you can file the paper for later reference. You can also combine any of these three options. For example, you can route it and mark it for filing after routing. But try not to put it aside to handle later. That just means you have to read and think about the same piece of paper twice. If you will have to get back to it later, stick a note on it defining what action is needed. That way you won't have to start all over when you pick it up again.

Organizing Your Files

Files are only useful if you can *find* the information you tuck away in them. It's easy to lose important information in a set of files. The key to retrieving information is to set up simple, logical file names. A file name may seem perfectly logical to you at the time you set it up, but later you can't remember where you filed it. This is how important information gets lost.

It often makes sense to file important paperwork under two different file names. Here are a couple of ways you might do this. If you have a lot of repeat business with a customer, you can file the contract under that customer's name and again under "Contracts." You might want to file a material order under a customer's name and also under "Open Orders" (until the order has been filled). With a redundant filing system, you're more likely to be able to find the paper you need when you need it. Just don't make too many files.

Another useful filing tool is the file index. It's a list of all the file names that you can quickly scan when you're not sure exactly what file name you're looking for. It can help you locate the correct file name for the information you're after. Your index may list files for Credits, Creditors, Contracts, Contractors, and so on. If this index is on a computer data base, it's easy to update and sort. You can also add margin notes, as necessary, to remind you about the contents of the files.

Using a Computer

A personal computer (PC) is another useful tool for managing paperwork. Like most tools, it has advantages and disadvantages.

Here are the disadvantages. The hardware is expensive. The software (programs) required to run the hardware can be expensive. It takes time to learn to use the equipment properly. When you first begin using a PC, it will take longer to do most tasks than if you did them manually. And when the computer is broken, the paperwork may not get done at all.

On the plus side, consider the following. Once you learn how to use it, a computer can handle paperwork more quickly and more efficiently. It can compile, file, sort, edit and print all types of data with ease.

Choosing the right hardware, choosing the right software and learning to use them takes time and effort. Here are some important points to remember when selecting your hardware and software.

Hardware: The key to selecting the right computer hardware for your business is to *research before you buy.* Stick to manufacturers who have been in business for a while. Talking to their salespeople isn't enough. Talk to their other customers as well. Read consumer publications on computers. In your computer research, here's what to look for:

- Equipment with enough capacity for your current needs and for your foreseeable future needs.
- Equipment that's reliable and can handle a wide variety of programs. Make sure the programs you need are available for the PC you want.
- Peripherals (printers and monitors) that are compatible with the PC you select. If you buy a dot matrix printer, choose one that can do graphics and prints with near-letter quality type.
- Manufacturers that guarantee quick, efficient equipment servicing.
- Instruction manuals that are easy to read and understand.
- Total cost, not just the base cost, including cables, the necessary display boards, modems and disk drives.

Some computer dealers will allow you to rent a computer and use it before buying. This lets you to explore the virtues and limitations of the equipment before you commit to a purchase. The more experience you have with computers, the easier it'll be to decide which one best suits your business needs.

Software: Buy software that meets the needs of your particular business. And before you buy, make sure the software does what the salesperson says it can do. Insist on a demonstration. You'll want to look at the following programs: word processing, spread sheets and data base. You may also want to look at graphics programs.

Word processing— A word processing program lets you write a letter, rearrange it, correct it, check the spelling of each word, and store it on disk, all before you print it. Since you can make all the changes you want *before* you print the document, you can be sure it will be printed without errors.

A word processing program also lets you to store form letters and contracts on a disk for future use. You can then modify the basic letter or contract to fit different situations. For instance, let's say you have a customer who owes you money. You can create and store a letter requesting that payment be made by a certain date. Now let's say another customer comes up with an overdue account. Use the same collection letter that you did for the first customer. Just change the name, address and payment due date before you print the letter. You can use this same procedure for contracts and other documents.

Spread sheets— This type of program lets you list, change and total columns and rows of figures. For example, use a spread sheet program to list delinquent accounts by name, showing the amount paid and the amount due for 30, 60, 90 or more days. The program will total all rows and columns of figures and show what percentage of delinquent accounts are paid or past due. As you update your payment records, the spread sheet program recalculates the amount owed by each client. That's important information. And the computer can do it quickly and conveniently with a spread sheet program.

A spread sheet program can also handle your payroll calculations. It can calculate deductions from each person's gross income, the category for each deduction, and then print the information on each pay stub. The program can also give you quarterly totals for each deduction category so you know how much money you owe the government. The spread sheet program can handle any data that uses numbers and requires calculations.

Data base— Use a data base program to store and retrieve information, usually data that doesn't use numbers or require calculations. For example, use a data base program to list all of your clients and important information about each, such as the client's address, telephone number and current job status. The program can sort this list by any category on the list. It can also take selective bits of information from the data and print just those items. For instance, the program can pull out just the customer name and address and use this information to print mailing labels.

Some of the newer (and more expensive) programs combine the functions of word processing, spread sheets and data base. They allow you to use data from all three sources and combine the data into one document. For example, you can write a letter that includes numeric calculations (spread sheet data) or nonnumeric information (data base).

Graphics programs: There are also specialized graphics programs that help you to create pictures and then add text to the pictures. That's handy for creating advertising pieces, bar charts and graphs. Some graphics programs have a number of pictures from which you can choose. They also let you add text in a number of different type styles and sizes. Figure 9-5 is a typical computer-generated flier.

If you have a color monitor and a color printer, you can create multicolored graphics with some of the programs that are available. With the right hardware and software, you're limited only by your imagination.

ADVERTISING

The purpose of advertising is to let people know about your company and the services you offer. You may be the best contractor around, but if no one knows about you, you won't get the business you deserve. Let's look at some ways you can promote your drywall contracting business.

Selecting a Method

There are many ways to advertise. Only a few will be right for your business. Most drywall contractors use one or more of the the following: Yellow Pages in the telephone book, handbills or fliers, business cards, personal letters, newspaper ads, outdoor signs, direct mail, radio and TV, and trade magazines. Let's look at the advantages and disadvantages of each of these.

Yellow Pages: Many people who are looking for a contractor start with the Yellow Pages. In addition to having your name listed, you may want to have a display ad that

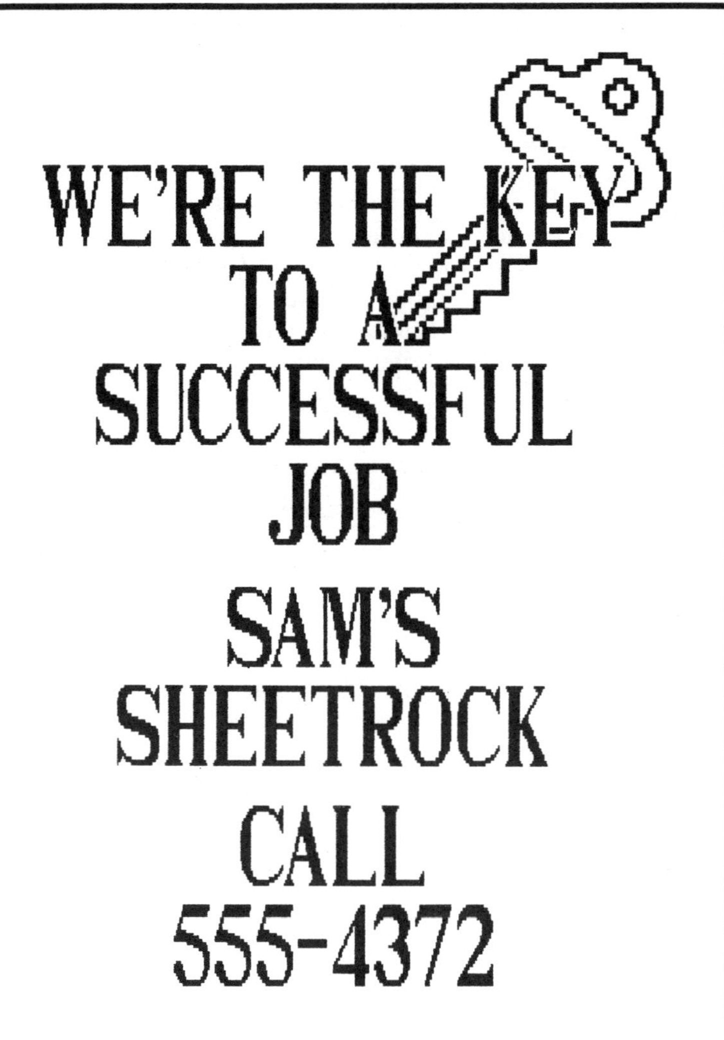

Computer-drawn flier
Figure 9-5

illustrates your services. And the more eye-catching it is, the better. You have to attract the reader's attention before you can communicate what you offer. You want to let the reader know what service you offer, and make him believe you're very good at that service.

Handbills or fliers: Distribute handbills and fliers to mailboxes in the area where you want to work. Random distribution may get people thinking about projects they'd forgotten but would like to get done. If your business includes remodeling work, be sure to send a flier to the homeowner who's just taken out a building permit. Figure 9-6 shows the simplest form of flier. It's easy, inexpensive — and it might bring in a few jobs. The computer-drawn flier in Figure 9-7 is a little more elaborate, but the message is the same: Call us for the job.

Business cards: This form of advertising is inexpensive, easy to carry with you, and to hand out to prospective customers. Your business card is probably your cheapest and most effective form of advertising.

Personal letters: Everyone likes to get mail. Sending a neat, informative letter to a list of homeowners will usually get results, if you select your list carefully. A computer and word processing program is ideal for sending personalized letters to a list. Always include a business card in the letter.

Newspaper ads: The advantage of newspaper ads is that they cover a wide area and reach a large number of people. Most of these people won't need your services at the moment, but they may be a source of future business. If they remember your ad, they may mention it to a friend who's looking for a contractor. I recommend using the classified section of the local paper, especially if you see ads placed by other construction trades. If these ads appear over and over, they're getting results. Your ads probably will too.

Outdoor signs: You don't need a billboard to attract customers. A small sign driven into the ground in front of any building where your crew is working will be enough. It tells the world that you're on the job in the neighborhood, doing quality work at a fair price.

CUSTOMER SATISFACTION
is our
GUARANTEE!

CALL Sam's Sheetrock

555-4372

- our schedule will match your schedule
- free estimates
- new construction and remodeling

Simple, inexpensive flier
Figure 9-6

A more elaborate flier
Figure 9-7

Designing Your Ads

All of your advertising has one goal: to attract a prospective customer's attention enough so that he or she contacts you. Until you get someone's attention, you can't communicate. You want the prospect to understand clearly the service you provide. And you also want the customer to know you're good at what you do. Here are four design elements that convey information likely to appeal to prospective customers: slogans, logos, color and shading, and type style.

Slogans: A slogan is a brief, attention-getting phrase. It's designed to grab someone's attention while conveying an important message about the service or product you offer. Famous slogans include: "We service what we sell," and "Ask the man who owns one." One electrical contractor has this slogan on the doors of his trucks: "Let us remove your shorts." That's a real attention-grabber. But it may not convey the right impression.

Logos: A logo is a symbol or picture used to identify your company. Your logo should relate to the work your company does. For example, you could use a drawing of a wide-blade taping knife for your logo. Or you could use a cartoon of a person taping drywall joints, applying plaster, or carrying half a dozen sheets of drywall into a building. A logo that's a symbol of what you do improves the chance that it will stick in someone's mind when they need drywall.

Color and shading: Use color and shading in your ad whenever you can. This helps attract attention. Be aware that certain colors may stimulate certain reactions in your readers. Red is a color that can stimulate strong emotion related to action, danger, heat and fear. Yellow is warm and inviting. Blue is a cool, quiet color. Green is a reminder of growth, life and springtime. Purple can be a reminder of royalty, wealth, mystery and dignity.

Use shading to provide emphasis, attract attention and give your ads a three-dimensional quality. In newspaper ads or other black and white advertising, shading can help your ad stand out from the rest. Highlighting the shading and surrounding it with blank space will also help emphasize your ad.

Type style: Use a type style that's easy to read. A reader isn't likely to spend time plodding through an ad that's cluttered with small print, unless there's an extremely provocative lead-in. Most readers just skim the ads. Use bold, contrasting type for the headline of your ad. Since the head is only a few lines long, you can get away with using an ornate type style. But make sure it's legible. And the information in the header has to be eye-catching. The header should grab the reader and stimulate him to read the rest of your ad.

The size of type is determined by the size of the ad. The height of type is measured in *points*. One point is 1/72 inch. The measurement is taken from the top to the bottom of the letter or number. The length of a line of type is measured in *picas*. A pica is 1/6 inch long. *Agate line* is a measure of space. One agate line is 1/14 inch high by one column wide. This information may be helpful when you're laying out an ad.

Always proofread your ads. Misspelled words in an ad make your company look sloppy. They convey the message that you're not very careful and don't pay much attention to detail. Protect your company against bad advertising. Always check your ads carefully.

Look at the advertising your competitors are doing. Is it the same or different from your advertising? Don't necessarily restrict yourself to the same media they're using. But keep an eye on what they're doing. It may give you some ideas about what type of advertising works well in a given area. And think about using advertising media that your competitors *aren't* using. You may attract some new customers that way.

The whole idea of advertising is to spread the word that you're in town, you have an important service to offer and you're good at what you do. You want people to remember your ads favorably. You want your company to come to mind when they need someone to install and finish drywall. Your company may not become a household word, but you want it to be known to as many people as possible.

WE COME TO THE END

We've covered a lot of ground in this manual. Maybe you didn't think hanging, taping and texturing drywall could be so complex. But making a good living in the construction industry is never easy. It requires knowledge, skill, hard work and some luck. I hope I've provided the information you need — the knowledge. Developing the skill, doing the hard work, and earning a few lucky breaks I'll have to leave to you.

GLOSSARY

A

Adhesive: Glue or mastic compound used to fasten gypsum wallboard to framing or to another wallboard panel.

Adhesive spreader: Notched board or trowel used to spread adhesive on the surface of a wallboard panel. The notches create ridges of adhesive with space between the ridges. This allows the adhesive to spread out when the panel is pressed against another surface.

All-purpose compound: A combined joint and topping compound. All-purpose compound doesn't bond quite as well as joint compound, but it's stronger than topping compound. You can use it for embedding joint tape. It also has the smooth-spreading quality of topping compound.

Back blocking: Piece of gypsum wallboard laminated to the backs of two gypsum panels at a joint between framing members.

Back clip: Clips attached to the back of gypsum wallboard. These clips fit into slots in the framing to hold the wallboard panels in place. These are used where wallboard panels are designed to be removable.

Back surface: The side of the panel that will be in contact with the framing members.

Backer board: Gypsum wallboard designed to be the first layer in a multilayer wall system or the base layer for applying acoustical ceiling tile.

Banjo: Automatic tool for applying joint tape alone, or joint tape and joint compound together.

Base coat: The first coat of compound when multiple coats are applied.

Batt: A precut piece of insulation designed to fit between framing members.

Batten: A strip used to cover joints.

Bead: A preformed strip used to reinforce corners or ends of wallboard panels. Bead also refers to the raised metal section at the corner of this strip.

Bed coat: The first coat of joint compound, into which the joint tape is embedded.

Beveled edge: The tapered factory edge of wallboard panels.

Binder: Additive mixed into the gypsum core to increase the bond between the paper and the core of the board. The binder is usually starch.

Bit: The portion of a screw gun that fits into the head of the screw.

Blanket: Insulation batts.

Bleeding: Discoloration that "bleeds" through to the surface. This usually occurs at a joint in the wallboard.

Blister: A portion of the facing paper or joint tape that comes unbonded from the surface of the panel.

Broad knife: Wide taping knife normally used for applying second and third coats of joint compound.

Buglehead screw: Gypsum wallboard screw with a flared head. This design helps seat the screw head slightly below the surface of the wallboard without tearing the facing paper.

Bundle: Two pieces of gypsum wallboard packaged face-to-face with a piece of paper tape along each short edge of the wallboard.

Butt joint: The joint formed when two pieces of wallboard are butted together.

Caging: Framing around pipes or other protrusions in a wall surface.

Calcine: To heat to a point that the water is released from the crystallized gypsum.

Caulk: To apply sealant to small gaps in a surface. Caulk is also the name of the compound used to seal small gaps. The compound is available in tubes or cartridges that fit a caulking gun.

Chalk box: Metal or plastic box that holds powdered chalk and string on a reel. The string is covered with chalk as it's withdrawn from the box. The chalk box is used to snap chalk lines on the wallboard surface.

Chalk line: Straight line formed by snapping a line covered with chalk powder.

Chase: Space through which piping or electrical service is run.

Chase wall: Wall around a pipe or electrical chase.

Cockle: Wrinkle or depression in the wallboard surface. The cockle occurs during manufacture.

Code: The marks on the back of gypsum wallboard panels that identify the type of board, the manufacturer, thickness and other specs.

Compressive strength: Strength of a material under a compressive (weight) load.

Control joint: Space left in a large expanse of wallboard to allow for expansion and contraction of the surface. The space is filled with a preformed metal joint that can be finished. The control joint allows the wall surface to move and prevents the wallboard from cracking and buckling.

Cooler nail: Gypsum drywall nail.

Core: The gypsum between the facing and backing papers of a wallboard panel.

Coreboard: The gypsum board designed to fit between wallboard panels in a self-supporting gypsum wall.

Corner bead: The preformed metal angle that is applied to outside corners of walls to reinforce the corner and protect the wallboard from impact damage.

Corner tool: Tool designed to apply joint compound simultaneously to both sides of an inside or an outside corner.

Crimping tool: Tool designed to clinch corner bead or metal studs into place.

Cripple: Short stud used between a header and a top plate or a sill and a sole plate in wall framing.

Crown: The buildup of joint compound over a taped joint.

Cupped taper: Defect in the tapered edge of a wallboard panel. The taper is dished so the edge of the taper is higher than the center of the taper.

Cut end: The cut edge of a gypsum wallboard panel where the gypsum core is exposed.

Daub: Spot of adhesive.

Dead load: The load on a building structure caused by the weight of the structure itself. The dead load on floor joists is the load caused by the flooring. The live load is the load caused by people, furniture and anything else that isn't part of the building.

Decibel: Standard measure of level of sound pressure.

Deflection: The distance a structure moves when a load is applied to it.

Deformation: The change in shape of a solid when a load is applied to it.

Delamination: Separation of plies that were fastened together with an adhesive bond.

Demountable partition: Temporary or moveable partition.

Dew point: Temperature at which moisture in the air condenses on a surface.

Dimple: Depression left in the surface of the wallboard when a nail is set by a hammer.

Double nailing: Wallboard fastening method using pairs of nails instead of single nails.

Dry taping: Applying joint tape with an adhesive other than joint compound.

Drywall: Gypsum wallboard.

E

Eased edge: Factory-tapered or rounded edge of a wallboard panel.

Edge: Refers to the long edge of a wallboard panel. The edge is normally covered with paper.

Embed: To apply joint tape to wet joint compound and cover it with another layer of joint compound.

End: The short edge of a wallboard panel. The panel end reveals the gypsum core.

Etched nails: Nails that are chemically treated to increase their holding power in wood.

F

Face: The surface of the wallboard panel that's exposed (away from the framing).

Face layer: Outer layer of wallboard when there are two or more plies.

Factory edge: Long edge of the wallboard panel. It normally comes covered with paper.

Feathered edge: The edge of a layer of joint compound that tapers smoothly to blend in with the surface of the wallboard.

Field: The surface area of a gypsum wallboard panel.

Fire blocking: Pieces of framing between the studs that run perpendicular to the studs. They prevent fire from spreading in the framing.

Fire taping: Taping the joints in gypsum wallboard to eliminate a path where fire might travel. This tape is used without a finish coat of joint compound.

Flanking path: Sound transmission path around a structure that is designed to stop or reduce sound transmission.

Floating angle: Applying wallboard without all of the fasteners in a corner. This allows for structural movement and reduces the possibility of the wallboard cracking.

Floating edge: Installation method where the long edge of the wallboard is not supported by a framing member.

Floating joint: Wallboard panel joint that is not located over a framing member.

Framing: The structural elements of a building.

Framing member: Individual part of the framing, such as a wall stud.

Furring: Wood or other material fastened to a surface before attaching the wallboard. Furring provides a gap between the wallboard and the building surface. It can also be used to make the surface plumb or level before attaching the wallboard.

Glue gun: Device used for applying adhesive from a cartridge.

Gray paper: Backing paper used on gypsum wallboard and the paper used on both sides of gypsum backing board.

Greenboard: Term generally used to refer to water-resistant gypsum wallboard.

Gusset: Wood or metal plate used to tie two structural members together and transfer a load between them.

Gypsum: Hydrated calcium sulfate.

Gypsum board: Wallboard that is manufactured with a gypsum core and paper or other facing.

Gypsum lath: Gypsum wallboard designed to be used as the backing for plaster.

Gypsum sheathing: Exterior gypsum wallboard used as the base for the finish building siding.

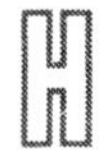

Hammer fracture: Tear in the wallboard facing paper caused by over-driving a nail.

Hard edge: A stronger section of core along the paper-bound edges of wallboard. It's designed to resist damage.

Hatchet: Hand tool with a blade on one end of the head and a nail-driving surface on the other end of the head.

Header: Structural member over a door or window. The header transfers upper structure loads to the studs alongside the door or window.

High joint: Wallboard joint that is raised above the rest of the surface of the wallboard.

Horizontal application: Installing gypsum wallboard with the long edges perpendicular to the framing members.

Humidity: Moisture in the air.

I

Insulation: Material designed to retard the transmission of heat or sound.

Isolation: Separation from a sound source.

J

Jamb: The casing around the inside of an opening.

Joint: Juncture between two pieces.

Joint compound: Putty-like material used to embed joint tape, cover joints and cover fastener heads in gypsum wallboard.

Joint photographing: When the joint or its shadow shows through the wallboard finish.

Joint tape: Special paper or fiberglass tape used to cover and reinforce joints in gypsum wallboard.

Joist: Horizontal structural member that supports the load of a floor or ceiling.

Knife: Tool used for cutting. A knife is also a tool used to spread joint or texture compound.

L

Laminated: Two or more layers joined with adhesive.

Lath: The backing to which plaster is applied.

Leaning edge: The factory taper on the long edge of a wallboard panel.

Live load: Temporary or movable load on a structure that isn't part of the building structure. This includes people, snow, furniture, etc.

Load-bearing partition: Wall that carries part of the load of the structure.

Loading pump: Hand pump used to load compound into application tools.

Manila paper: Facing paper for standard gypsum wallboard.

Mastic: Adhesive.

Mesh tape: Fiberglass joint tape.

Metal studs: Galvanized steel framing members that are preformed out of sheet steel.

Mildew: Fungus growth that is encouraged by dampness.

Miter joint: Joint that is cut at an angle, usually 45 degrees.

Mud: Joint compound.

Mud pan: Rectangular container that holds the joint compound for you during application. The pan is long enough so you can use a wide taping knife to take the compound out of the pan.

Multilayer: Two or more layers.

N

Nail pops: When nail heads "pop" through the surface of the wallboard.

Nail spotter: An applicator for applying joint compound to fastener head depressions.

Non-load-bearing: Wall or partition that doesn't carry any of the load of the structure. It serves only as a divider.

On center (o.c.): This refers to the centerline spacing of items. The "on center" distance between items is the distance measured from the center of one item to the center of the next.

Out of square: When true 90-degree angles do not exist but should exist.

Parallel application: Installing gypsum wallboard with the long edges parallel to the framing members.

Partition system: An assembly used as a divider between rooms.

Party wall: Wall that is common between two different living areas.

Perlite: Lightweight expanded mineral product.

Pilaster: Square building column that is part of a wall and projects out from the surface of the wall.

Plate: Horizontal framing member, attached to the studs, that forms the top or bottom of a wall.

Plumb: Vertical.

Primer: Specially formulated paint used as the first coat to seal a surface and provide a base for a final coat of paint.

Quick-set: A term referring to a fast-setting compound.

Racking: Sideward force tending to move a structure out of plumb or out of alignment.

Radiant heat: Heating by transmitting energy through space.

Rasp: A coarse file.

Ready-mix: Factory-mixed.

Relative humidity: A measure of the amount of moisture in the air.

Resilient channel: Formed sheet-metal section installed between the wallboard and the framing to reduce sound transmission through walls.

Ridging: Wallboard defect consisting of a raised line along the surface of a finished joint.

Ring-shank: Nails with rings around the shanks. This increases the gripping power of the nails.

Rip: Generally used to refer to a lengthwise cut. In the case of wood, it means to cut with the grain of the wood.

Ripper: Narrow strips of wallboard.

Runner: Tracks or strips placed along ceiling and floor, usually for an attachment point.

Sanding: Smoothing with sandpaper or other fine abrasive.

Sealer: Specially formulated paint used to seal a surface and provide a base coat for a final coat of paint.

Seam: A taped wallboard joint.

Serrated: Saw-tooth edge.

Score: To scribe a line along a surface using a knife.

Screw gun: Power screwdriver that usually has a clutch, magnetic screwdriver bit and adjustable screw depth.

Screws: Fasteners with helical ridges along the shank.

Scribe: To mark a line along a surface.

Scuff: To scrape the surface, leaving surface irregularities.

Shadowing: Visible joint tape edges in a finished wallboard joint.

Shaft wall: The wall of an enclosure built to separate stairways, elevators, wiring and plumbing.

Shear: To cut off at right angles. This term also refers to resistance to sideways movement.

Sheathing: Outer covering.

Shim: To use thin pieces of material to level or plumb a structure. The term also refers to the material used for this purpose.

Sizing: A surface sealant.

Skim coat: A thin coat.

Skip-trowel: A texturing technique.

Soffit: A covering over a space under the eaves of a structure or over cabinets.

Sound transmission: The ability of a gas, liquid or solid to conduct sound.

Span: Distance between supports.

Splice: Join together.

Spotting: Covering fastener heads with joint compound.

Staples: U-shaped fasteners.

Stud: Vertical framing member.

Stud finder: Device used to locate studs in a wall surface.

T

Taper: Sloped reduction in thickness.

Taping tools: Tools designed for the application of tape and compound and for finishing joints in gypsum wallboard.

Texture: Surface decoration.

Tongue-and-groove: An edge shaped to provide an interlocking joint.

Topping compound: Smooth compound used for the final coat over wallboard joints and fastener heads.

Trowel: Hand tool with a handle and a rectangular flat metal surface.

Type X gypsum wallboard: Fire-resistant wallboard.

U

Utility knife: Knife with a short, replaceable blade.

V

Veneer: Thin covering.

Veneer-base: Special type of gypsum wallboard used as the base for veneer plaster.

Veneer plaster: Special plaster you can apply in a thin coating. Hardens to an abrasion-resistant surface.

Vertical application: Installing wallboard panels with the long edges parallel to the studs.

Void: Hollow space in the core of gypsum wallboard.

Water-resistant: Material that minimizes the absorption of water.

Wet-sanding: Smoothing with the aid of water.

The estimating forms on the following pages and the contract on pages 189 and 190 are reproduced here with the permission of the publisher. Full-size (8½ by 11) copies of these forms suitable for reproduction on your copy machine or by your instant printer are available at no charge with any order from Craftsman Book Company, 6058 Corte del Cedro, Carlsbad, CA 92009. Ask for the 16-page *Builders & Estimators Form Book* or use the order form that begins on page 286 of this manual.

Material Estimate

(transfer Quantities to Labor Estimate)
(transfer Quantities and Material Costs to Detailed Cost Estimate)

	Initials	Date
Prepared By		
Approved By		

Project: ______________________

Location: ______________________

Estimate section: ______________ **No.** ________

Date: ______________

Page ______ **of** ______

Item					Material Cost		Suggested Source/Notes
No.	Description	Dimensions	Quantity	Unit	Unit Cost	Total Cost	
	Total						

Labor Estimate

(transfer line totals to Detailed Cost Estimate)
(transfer Equipment to Equipment Estimate)

	Initials	Date
Prepared By		
Approved By		

Project: ____________________

Location: ____________________

Estimate section:____________ **No.**________

Date: ____________

Page ______ **of** ______

Item				Labor Requirements		Labor Cost		Equipment/ Notes
No.	Description	Quantity	Unit	Man-Hours/ Unit	Total Man-Hours	Unit Cost	Total Cost	
	Total							

Detailed Cost Estimate

(transfer sheet totals to Estimate Summary)

	Initials	Date
Prepared By		
Approved By		

Project: ______

Location: ______

Estimate section: ______ No. ______

Date: ______

Page ______ of ______

Item			Material		Labor		Equipment		Total		Actual Job Cost	
Description	Quantity	Unit	Unit Cost	Total	Unit Cost	Total	Unit Cost	Total	Unit Cost	Total	Unit Cost	Total
Totals →												

Estimate Summary

Project: ____________ Location: ____________ Estimate prepared for: ____________ Estimate prepared by: ____________ Phone: ____________	Total estimate cost: Total building SF floor area: Total cost per SF floor area:	Date of estimate: ____________ Date projected to: ____________ Page ________ of ________

Section No.	Estimate Section	Estimated Quantity	Unit	Unit Cost	Total Cost

INDEX

Mail This No Risk Card Today

☐ **Please send me the books checked for 10 days free examination. At the end of that time I will pay in full plus postage (& 6% in Calif.) or return the books postpaid and owe nothing.**

☐ **Enclosed is my full payment or Visa/MasterCard/American Express number. Please rush me the books without charging for postage.**

☐ 19.75 Carpentry for Residential Construction
☐ 19.00 Carpentry in Commercial Construction
☐ 24.25 Contractor's Guide to the Building Code Rev.
☐ 17.00 Estimating Home Building Costs
☐ 28.00 Estimating Painting Costs
☐ 19.50 National Construction Estimator
☐ 19.25 Paint Contractor's Manual
☐ 21.25 Painter's Handbook
☐ 21.00 Running Your Remodeling Business
☐ 14.25 Wood-Frame House Construction

In a hurry?
We accept phone orders charged to your MasterCard, Visa or American Express.
Call (619) 438-7828

☐ **Visa** ☐ **MasterCard** ☐ **American Express**

Name (Please print clearly)

Company

Address

City/State/Zip

Expiration date ____________________

Card # ____________________

NO POSTAGE
NECESSARY
IF MAILED
IN THE
UNITED STATES

BUSINESS REPLY MAIL

FIRST CLASS MAIL PERMIT NO.271 CARLSBAD, CA

POSTAGE WILL BE PAID BY ADDRESSEE

Craftsman Book Company
6058 Corte Del Cedro
P. O. Box 6500
Carlsbad, CA 92008-0992

NO POSTAGE
NECESSARY
IF MAILED
IN THE
UNITED STATES

BUSINESS REPLY MAIL

FIRST CLASS MAIL PERMIT NO.271 CARLSBAD, CA

POSTAGE WILL BE PAID BY ADDRESSEE

Craftsman Book Company
6058 Corte Del Cedro
P. O. Box 6500
Carlsbad, CA 92008-0992